Learning to Predict Climate Variations Associated with El Niño and the Southern Oscillation

Accomplishments and Legacies of the TOGA Program

Advisory Panel for the Tropical Oceans and Global Atmosphere Program
(TOGA Panel)

Climate Research Committee

Board on Atmospheric Sciences and Climate

Commission on Geosciences, Environment, and Resources

National Research Council

NATIONAL ACADEMY PRESS
Washington, D.C. 1996

NOTICE: The project that is the subject of this report was approved by the Governing Board of the National Research Council, whose members are drawn from the councils of the National Academy of Sciences, the National Academy of Engineering, and the Institute of Medicine. The members of the committee responsible for the report were chosen for their special competences and with regard for appropriate balance.

This report has been reviewed by a group other than the authors according to procedures approved by a Report Review Committee consisting of members of the National Academy of Sciences, the National Academy of Engineering, and the Institute of Medicine.

This work is funded in part by contract CMRC 50-DKNA-5-00015/C from the National Oceanic and Atmospheric Administration (NOAA) and printed with funds from U.S. Department of Agriculture grant 403K06506534. The views expressed herein are those of the authors and do not necessarily reflect the views of NOAA or any of its sub-agencies.

Library of Congress Catalog Card Number 96-70442
International Standard Book Number 0-309-05342-0

Additional copies of this report are available from:
National Academy Press
2101 Constitution Ave., NW
Box 285
Washington, DC 20055
800-624-6242
202-334-3313 (in the Washington metropolitan area)

COVER ART: Reproduction of the batik "Sunday Morning" by Susan Schneider, from the collection of Edward Sarachik. Ms. Schneider runs her own batik card business in Seattle, Washington. Her father, Harry Wexler, was Chief Scientist of the U.S. Weather Bureau. Her uncle, Jerome Namais (member of the National Academy of Sciences), was a major developer of long-range weather prediction and a pioneer in the study of short-term climate variations. "Sunday Morning" is based on a black and white photograph from the 1890s of residents in the town of Oak Bluffs, Massachusetts, heading for church. The TOGA Panel selected it for the cover to evoke the beginnings of a new age of climate prediction.

Copyright 1996 by the National Academy of Sciences. All rights reserved.
Printed in the United States of America.

ADVISORY PANEL FOR THE
TROPICAL OCEANS AND GLOBAL ATMOSPHERE PROGRAM
(TOGA PANEL)

EDWARD S. SARACHIK (*Chair*), University of Washington, Seattle
ANTONIO J. BUSALACCHI, NASA Goddard Space Flight Center, Greenbelt, Maryland
ROBERT E. DICKINSON, University of Arizona, Tucson
STEVEN ESBENSEN, Oregon State University, Corvallis
DAVID HALPERN, Jet Propulsion Laboratory, California Institute of Technology, Pasadena
DENNIS L. HARTMANN, University of Washington, Seattle
ROBERT A. KNOX, Scripps Institution of Oceanography, La Jolla, California
ANTS LEETMAA, National Oceanic and Atmospheric Administration, Camp Springs, Maryland
ROGER LUKAS, University of Hawaii, Honolulu
STEPHEN E. ZEBIAK, Lamont-Doherty Earth Observatory, Palisades, New York

Staff

MARK D. HANDEL, Senior Program Officer
ELLEN F. RICE, Reports Officer
THERESA M. FISHER, Administrative Assistant
MARK BOEDO, Project Assistant

CLIMATE RESEARCH COMMITTEE

ERIC J. BARRON (*Chair*), Pennsylvania State University, University Park
DAVID S. BATTISTI, University of Washington, Seattle
RUSS E. DAVIS, Scripps Institution of Oceanography, La Jolla, California
ROBERT E. DICKINSON, University of Arizona, Tucson
THOMAS R. KARL, National Climatic Data Center, Asheville, North Carolina
JEFFREY T. KIEHL, National Center for Atmospheric Research, Boulder, Colorado
CLAIRE L. PARKINSON, NASA Goddard Space Flight Center, Greenbelt, Maryland
STEVEN W. RUNNING, University of Montana, Missoula
KARL E. TAYLOR, Lawrence Livermore National Laboratory, California

Ex Officio Members

DOUGLAS G. MARTINSON, Lamont-Doherty Earth Observatory, Palisades, New York
EDWARD S. SARACHIK, University of Washington, Seattle
SOROOSH SOROOSHIAN, University of Arizona, Tucson
PETER J. WEBSTER, University of Colorado, Boulder
W. LAWRENCE GATES, Lawrence Livermore National Laboratory, California

Staff

WILLIAM A. SPRIGG, Director
MARK D. HANDEL, Senior Program Officer
DORIS BOUADJEMI, Administrative Assistant
KELLY NORSINGLE, Senior Project Assistant

BOARD ON ATMOSPHERIC SCIENCES AND CLIMATE

JOHN A. DUTTON (*Chair*), Pennsylvania State University, University Park
ERIC J. BARRON, Pennsylvania State University, University Park
WILLIAM L. CHAMEIDES, Georgia Institute of Technology, Atlanta
CRAIG E. DORMAN, consultant, Arlington, Virginia
FRANCO EINAUDI, NASA Goddard Space Flight Center, Greenbelt, Maryland
MARVIN A. GELLER, State University of New York, Stony Brook
PETER V. HOBBS, University of Washington, Seattle
WITOLD F. KRAJEWSKI, The University of Iowa, Iowa City
MARGARET A. LeMONE, National Center for Atmospheric Research, Boulder, Colorado
DOUGLAS K. LILLY, University of Oklahoma, Norman
RICHARD S. LINDZEN, Massachusetts Institute of Technology, Cambridge
GERALD R. NORTH, Texas A&M University, College Station
EUGENE M. RASMUSSON, University of Maryland, College Park
ROBERT J. SERAFIN, National Center for Atmospheric Research, Boulder, Colorado

Staff

WILLIAM A. SPRIGG, Director
H. FRANK EDEN, Senior Program Officer
MARK D. HANDEL, Senior Program Officer
DAVID H. SLADE, Senior Program Officer
ELLEN F. RICE, Reports Officer
DORIS BOUADJEMI, Administrative Assistant
KELLY NORSINGLE, Senior Project Assistant

COMMISSION ON GEOSCIENCES, ENVIRONMENT, AND RESOURCES

GEORGE M. HORNBERGER (Chair), University of Virginia, Charlottesville
PATRICK R. ATKINS, Aluminum Company of America, Pittsburgh, Pennsylvania
JAMES P. BRUCE, Canadian Climate Program Board, Ottawa, Ontario
WILLIAM L. FISHER, University of Texas, Austin
JERRY F. FRANKLIN, University of Washington, Seattle
DEBRA KNOPMAN, Progressive Foundation, Washington, D.C.
PERRY L. McCARTY, Stanford University, California
JUDITH E. McDOWELL, Woods Hole Oceanographic Institution, Massachusetts
S. GEORGE PHILANDER, Princeton University, New Jersey
RAYMOND A. PRICE, Queen's University at Kingston, Ontario
THOMAS C. SCHELLING, University of Maryland, College Park
ELLEN SILBERGELD, University of Maryland Medical School, Baltimore
VICTORIA J. TSCHINKEL, Landers and Parsons, Tallahassee, Florida

Staff

STEPHEN RATTIEN, Executive Director
STEPHEN D. PARKER, Associate Executive Director
MORGAN GOPNIK, Assistant Executive Director
GREGORY SYMMES, Reports Officer
JAMES MALLORY, Administrative Officer
SANDI FITZPATRICK, Administrative Associate
MARQUITA SMITH, PC Analyst

The National Academy of Sciences is a private, nonprofit, self-perpetuating society of distinguished scholars engaged in scientific and engineering research, dedicated to the furtherance of science and technology and to their use for the general welfare. Upon the authority of the charter granted to it by the Congress in 1863, the Academy has a mandate that requires it to advise the federal government on scientific and technical matters. Dr. Bruce M. Alberts is president of the National Academy of Sciences.

The National Academy of Engineering was established in 1964, under the charter of the National Academy of Sciences, as a parallel organization of outstanding engineers. It is autonomous in its administration and in the selection of its members, sharing with the National Academy of Sciences the responsibility for advising the federal government. The National Academy of Engineering also sponsors engineering programs aimed at meeting national needs, encourages education and research, and recognizes the superior achievements of engineers. Dr. William A. Wulf is interim president of the National Academy of Engineering.

The Institute of Medicine was established in 1970 by the National Academy of Sciences to secure the services of eminent members of appropriate professions in the examination of policy matters pertaining to the health of the public. The Institute acts under the responsibility given to the National Academy of Sciences by its congressional charter to be an adviser to the federal government and, upon its own initiative, to identify issues of medical care, research, and education. Dr. Kenneth Shine is president of the Institute of Medicine.

The National Research Council (NRC) was organized by the National Academy of Sciences in 1916 to associate the broad community of science and technology with the Academy's purposes of furthering knowledge and advising the federal government. Functioning in accordance with general policies determined by the Academy, the Council has become the principal operating agency of both the National Academy of Sciences and the National Academy of Engineering in providing services to the government, the public, and the scientific and engineering communities. The Council is administered jointly by both Academies and the Institute of Medicine. Dr. Bruce M. Alberts and Dr. William A. Wulf are chairman and interim vice chairman, respectively, of the National Research Council.

PREFACE

El Niño and the Southern Oscillation, collectively called ENSO, are primary drivers of interannual climate variability. Prior to the initiation of the Tropical Oceans and Global Atmosphere (TOGA) Program in 1985, scientists had only the beginnings of a picture of ENSO. They had scant, scattered means of observing it. Ocean observations, in particular, were usually available after considerable delay. Scientists had only begun to understand the coupled dynamics of the ocean and atmosphere system; most previous research had concentrated on either the oceanic or the atmospheric response to a specified forcing by the other fluid. In 1982, the onset of the largest ENSO warm event (El Niño) of this century was not even recognized, let alone predicted, while in its early stage of development because available observations were so meager, aerosols from the eruption of El Chichon had contaminated satellite-based observations, means of effective data dissemination and interpretation were so lacking, and predictive models were so rudimentary. Research efforts at that time were still strongly divided between meteorologists and oceanographers, and the disciplinary barriers between them were significant. Useful ENSO predictions were a distant dream.

The TOGA program, which began 1 January 1985 and ended 31 December 1994, addressed the important challenges of ENSO prediction. As the TOGA decade closed, the improvement in our understanding of short-term climatic fluctuations on our planet was clearly evident. A substantial network for oceanic and atmospheric observations is in place in the tropical Pacific and, to a lesser extent, in the other tropical oceans. Most of the data from that network are transmitted immediately and are accessible to researchers shortly after collection. Models have improved in quality. Regular predictions of aspects of ENSO are now made by a number of groups worldwide. Currently, these predictions are significantly better than predictions based on climatology for only limited geographic regions. Even in regions where skill is greatest, forecasts still fail. Nevertheless, the predictions are demonstrably skillful, enough so that they are taken seriously and used to guide national economic strategies and choices in some of the countries most affected by ENSO.

International efforts to institutionalize short-term climate predictions for practical, regionally tailored ends are taking shape and expanding. An international community of scientists with firm roots in both meteorology and oceanography has come of age, learned how to talk to each other, and joined forces to carry out major cooperative research efforts. TOGA, in many respects, now represents a model for international and interdisciplinary scientific research.

In view of such success, holding to the end date of 1994 for TOGA that was scheduled somewhat arbitrarily in 1985 may seem odd or rigid. The decision to do so, however, was not made casually. It necessitated thought about how to sustain certain activities, such as the observing system, that were nourished by TOGA but that must now remain healthy under other stewardship. However, formally ending TOGA has also ignited a valuable, energetic reassessment of the state of the art in understanding and predicting all aspects of seasonal-to-interannual climate variations. This assessment has involved a broad community of scientists. The initial planning for the new Global Ocean–Atmosphere–Land System (GOALS) program, an ambitious attempt to extend our knowledge of ENSO and other short-term variations of climate, has flowed from this reassessment (see NRC 1994b). New scientists have joined the process of shaping GOALS, bringing new ideas and fresh enthusiasm. If the end of TOGA has been unsettling and worrisome, it has been reinvigorating as well.

The National Research Council's Advisory Panel for the Tropical Oceans and Global Atmosphere Program has played an active and very important role in shaping and guiding the U.S. contributions to TOGA. The panel as a group as well as some of its members as individuals, have been closely associated with the program. The Climate Research Committee, as the "parent" of the TOGA Panel, has asked the Panel to write a retrospective on TOGA, its accomplishments, and its shortcomings, with an emphasis on the U.S. contributions. Our hope is that this report, in conjunction with the original planning reports (NRC 1983, 1986) and mid-term assessments of scientific progress (NRC 1990), will provide a valuable milestone on the scientific path from TOGA, through GOALS, into future efforts to comprehend and cope with Earth's climatic machinery.

                                              Eric Barron, Chair
                                              Climate Research Committee

ACKNOWLEDGMENTS

The TOGA Panel would like to thank David Battisti, Grant Branstator, Inez Fung, Richard Gammon, Michael Halpert, D. Edmunds Harrison, Mojib Latif, Gary Mitchum, James Moum, John Marsh, Michael McPhaden, Robert Molinari, James W. Murray, Peter Niiler, Kevin Trenberth, Peter Webster, and Yuan Zhang for contributions to this report, and also the staff of the Board on Atmospheric Sciences and Climate, especially Mark Handel, for their diligence in preparing this report for publication. The current members of the panel are grateful to the many previous members (listed in Appendix A), who helped design the TOGA Program and provided part of the foundation for this report.

> Edward Sarachik, Chair
> TOGA Panel

CONTENTS

SUMMARY 1

1. INTRODUCTION 5
   El Niño and the Southern Oscillation (ENSO) 7
   Concept of the TOGA Program 8
   Purpose of this Report 9

2. GROWTH OF THE TOGA PROGRAM 12
   ENSO: A Coupled Ocean–Atmosphere Phenomenon 12
   Emergence of a Coherent Effort (1970–1984) 14
   Development of the TOGA Program 18
   Scientific Plan for TOGA (1985) 19

3. COMPONENTS OF THE U.S. TOGA PROGRAM 22
   Observations of ENSO 24
   Process Studies 50
   Modeling ENSO 66
   Prediction 71
   TOGA Products 76
   Problems and Shortcomings 78

4. WHAT WE'VE LEARNED 81
   Observations of ENSO in the Tropical Pacific 82
   Effects of ENSO on the Rest of the Globe 87
   Theories of ENSO 96
   Working in a Larger Community 106

5. ORGANIZATION 110
   U.S. Organizational Arrangements 111
   International Organizational Arrangements 117

6. APPLICATIONS OF ENSO PREDICTION 123
   Development of Applications and Assessments 124
   Applications of Regional ENSO Forecasts 127

7. THE FUTURE 132
   What TOGA Didn't Do 133
   Obstacles to Progress 135
   An International Research Institute for Climate Prediction (IRICP) 138
   GOALS and CLIVAR 139

REFERENCES 141
APPENDICES 165
    A. Members of the TOGA Panel..165
    B. TOGA Products 166
    C. Acronyms and Other Abbreviations...169

LIST OF FIGURES

Figure 1. Schematic of ENSO...6
Figure 2. Correlations of annual-mean sea-level pressure with the pressure at Darwin, Australia...7
Figure 3. The major conceptual components of TOGA 22
Figure 4. Funding for U.S. TOGA 23
Figure 5. The TOGA *In Situ* Pacific Basin Observing System...................26–27
Figure 6. An ATLAS (Autonomous Temperature Line Acquisition System) Mooring 38
Figure 7. Climatology of the near-surface equatorial ocean at 110°W.............44
Figure 8. Composite structure of the intensive operation period for TOGA COARE...51
Figure 9. Monthly-mean anomalies of sea surface temperature averaged over 5°S to 5°N during the TOGA years 54
Figure 10. Correlations for predictions of anomalies of equatorial sea surface temperature using fully coupled atmosphere–ocean models 74
Figure 11. Evolution of sea surface temperature in the tropical Pacific 84
Figure 12. Total field and anomalies of sea surface temperature for January 1992 85
Figure 13. Analyses of interannual and lower frequency variations of sea surface temperature throughout the entire Pacific 86
Figure 14. Organization of TOGA within the United States 111
Figure 15. International organization for TOGA 118

LIST OF TABLES

Table 1. TOGA data requirements..28
Table 2. Rain conditions for India and associated phase of ENSO 91

Learning to Predict Climate Variations Associated with El Niño and the Southern Oscillation

Accomplishments and Legacies of the TOGA Program

SUMMARY

The Tropical Oceans and Global Atmosphere (TOGA) Program was designed:
1. To gain a description of the tropical oceans and the global atmosphere as a time-dependent system, in order to determine the extent to which this system is predictable on time scales of months to years, and to understand the mechanisms and processes underlying that predictability;
2. To study the feasibility of modeling the coupled ocean–atmosphere system for the purpose of predicting its variations on time scales of months to years; [and]
3. To provide scientific background for designing an observing and data transmission system for operational prediction if this capability is demonstrated by the coupled ocean–atmosphere system.

These objectives were formulated because of the recognized scientific importance of natural variability in the climate system, the consequences of the variations for the economies and societies of the world, and the potential value of skillful predictions. The United States cooperated with the international community, as part of the World Climate Research Programme, to achieve these objectives. This document examines the extent to which the TOGA Program succeeded in meeting these objectives and the role of U.S. participation in the program. However, it has been difficult to determine which activities were part of, or initiated because of, TOGA. This difficulty reflects the ability of U.S. TOGA to leverage its resources by integrating its efforts with other national and international climate-research activities. This report takes an expansive view by considering almost all research on seasonal-to-interannual climate research from 1985 through 1994 to be associated with the international TOGA program.

TOGA largely fulfilled, and in some ways exceeded, its objectives. The program built and maintained the TOGA Observing System, which provided observations of El Niño and the Southern Oscillation (ENSO) in the previously poorly sampled region of the tropical Pacific. It developed coupled atmosphere–ocean models of the tropical Pacific Ocean, some of which now demonstrate skill in the prediction of tropical sea surface temperature months to a year or so in advance. TOGA spawned several process experiments, especially the Coupled Ocean–Atmosphere Response Experiment (COARE), which explored aspects of tropical atmosphere–ocean coupling that were poorly understood. As a result of these activities, a new capability for observing the surface and near-surface ocean and atmosphere in real time has been developed, a new depth of understanding of ENSO has been achieved, theoretical models for the mechanism of ENSO have been proposed, and the beginning of an understanding of the relationship between tropical sea-surface-temperature perturbations and the atmospheric response in the middle latitudes has been attained.

However, TOGA failed to meet its objectives fully. The program did not live up to its name by completing studies throughout the tropical oceans and the global atmosphere, but instead concentrated only on the large signal of seasonal-to-interannual coupled atmosphere–ocean interactions in the tropical Pacific (the ENSO phenomenon). Variability and predictability arising from processes in other ocean basins, and from the interactions of the atmosphere with land and with ice, received scant attention. Understanding of the effects of tropical sea surface temperatures on the higher latitudes developed slowly. The program did not address the possibility of variability and predictability on seasonal-to-interannual time scales arising from interactions within the middle latitudes. Research on seasonal-to-interannual variations of atmospheric circulation conducted under other auspices was not well integrated with the TOGA Program. Implementation of both the COARE field program, the largest process study in TOGA, and the Tropical Atmosphere/Ocean (TAO) array, the heart of the TOGA Observing System in the tropical Pacific region, were both accomplished only near the end of TOGA, so that their usefulness for improving prediction could not be demonstrated within the time frame of the TOGA Program itself. The expected satellite-based observation system, especially a scatterometer for estimating surface winds, did not materialize. Observational strategies for understanding seasonal-to-interannual variability were not developed for anywhere but the tropical Pacific. These failures, combined with the unexpected difficulties in coupling complex atmosphere and ocean models, meant that while much progress was made, the full accomplishment of the TOGA objectives will be realized only by future programs.

By developing short-term climate predictions (predicting tropical Pacific sea surface temperatures months to a year or so in advance), TOGA has transcended the confines of a research program. Climate prediction (and the applications of climate prediction) has become a motivation for limited-duration research programs, and will continue to demand attention far into the future. Tension developed during the second half of TOGA between research motivated by the ideal of operational prediction and actual operational prediction efforts. Incipient applications of climate predictions were initiated under TOGA and a prototype International Research Institute for Climate Prediction was designed. Such an institute would support the needs of individual regions, through Regional Applications Centers. Peru, Brazil, and Australia are already finding applications of short-term climate predictions useful to their own economies and societies. Applications for the United States will depend on the development of predictive skill at higher latitudes; such skill is still meager.

The implementation of U.S. participation in TOGA was accomplished through the U.S. Global Change Research Program (USGCRP) and was aided by an unusual set of institutional arrangements. In the United States, the program was funded and administered from an interagency project office, and scientific advice was provided by the National Research Council (NRC) through its TOGA Panel. Internationally, the program was administered by the International TOGA Project Office, advised by the TOGA Scientific Steering Group, and resources were gathered by the Intergovernmental TOGA Board. These management arrangements grew organically, according to the needs of the program, and were largely responsible for its smooth functioning. The successful mix of science and applications demonstrated by TOGA can serve as a model for other components of the USGCRP.

TOGA involved studies of both the atmosphere and the ocean, as well as their mutual interaction. The program therefore had major effects on practice and education in the atmospheric and oceanic sciences. Meteorologists and oceanographers worked together on a single problem, crossing the disciplinary and institutional barriers in universities and government. The real-time distribution of oceanographic data, coincident with the growth of the information superhighway, made data freely available in ways totally different from those previously possible. Applications of short-term climate prediction established a link between scientists and society, translating society's concern with crop planning, water management, fisheries, disaster planning, etc. into a focused scientific endeavor.

Predictions of ENSO are not possible without observations to initialize prediction schemes in the region of the Pacific where ENSO is strongest. Research on ENSO also places a high priority in data from this region. It is essential to maintain what we already have, in particular, the upper-air observing network, satellite altimetry, and the upper-ocean and surface-meteorological measurements made in and over the ocean. Guidance provided in NRC 1994a remains relevant, and we paraphrase two of its recommendations here. The TOGA Observing System in the Pacific, especially the moorings that constitute the TAO array, should be continued because of its value for initializing and evaluating systems making predictions of seasonal-to-interannual climate variations. The components of the system should be maintained until a serious study of their impact on prediction reveals them to be of marginal value or until a more cost-effective technique is demonstrably ready to replace them.

The prediction of short-term climate variations is a nascent field. There is much to be learned on how to provide and apply such predictions. The provision and application of seasonal-to-interannual forecasts will require forging links between the climate research community and the communities of applied climatologists, social scientists, and users of climate information. TOGA has brought us to a time for the establishment of a prototype international institute for making predictions and demonstrating the applicability of these predictions, as previously recommended in NRC 1995b.

TOGA concentrated on only the strongest climate variation on seasonal-to-interannual time scales, ENSO, and concentrated on ENSO only in a limited geographic region where the signal is strongest, the tropical Pacific. Much research is still required to develop skill for predicting short-term climate variations caused by other processes or in other places. Unless they coordinate their efforts, researchers are unlikely to develop this skill efficiently. National and international programs are needed for research on, and development of, the exploitation of predictability and the making of predictions of seasonal-to-interannual climate variations throughout the world. In particular, the World Climate Research Programme study CLIVAR (Study of Climate Variability and Predictability)/GOALS (WCRP 1995) and its U.S. contribution GOALS, proposed in reports from the NRC (1994b, 1995a), provide a path for furthering the accomplishments and building on the legacy of TOGA.

1. INTRODUCTION

El Niño is an extensive warming of the upper ocean in the tropical eastern Pacific lasting three or more seasons. The Southern Oscillation is a widespread interannual oscillation in sea-level atmospheric pressure between one region near northern Australia and one in the central Pacific. These related phenomena, together called ENSO, are the largest short-term climate variations. The TOGA Program was designed to study seasonal-to-interannual variability and predictability around the globe, with an emphasis on ENSO. Available resources limited the program largely to studies of ENSO in the tropical Pacific region. Before the TOGA Program, observing systems were inadequate to recognize the warm phase of ENSO— associated with El Niño—until it was well underway. At the conclusion of the program, a tropical Pacific observing system had been established and predictions of ENSO were being made with some skill. This report reviews the accomplishments and legacies of the TOGA Program, and then makes recommendations for further progress in predicting seasonal-to-interannual variations in climate around the globe.

The climate of a location refers to more than just the local long-term average of the annual cycles of temperature and precipitation. It includes other quantities, such as winds, humidity, and cloudiness. It includes places other than the surface, such as the upper atmosphere and the oceans. It includes other surface conditions, such as the presence of snow and ice (glaciers and sea ice), soil moisture, and vegetation cover. Furthermore, climate information now includes the variability of all these conditions, such as the frequency of storms, the range and frequency of extreme conditions, and the variations from year to year.

Since the late eighteenth century, it has been accepted that the climate of the earth has undergone dramatic changes, but until recently it was thought that those changes occurred over periods of thousands of years and longer. We now recognize that climate varies on all time scales. This recognition has led to a transformation, as well as to confusion, in how we use the term "climate". For lack of a better term, we will consider a measure of the current state of the climate system to be an average over a period that includes several weather systems, i.e., an average over many weeks or longer. Climate variations, also called anomalies, are differences in the state of the climate system from normal conditions (averaged over many years) for that time of year. Short-term climate variations are periods of a few months up through a few years that are unusually warm or cool (or wet or dry). The TOGA (Tropical Oceans and Global Atmosphere) Program was the first organized effort to study, understand, and predict the year-to-year variations of the climate system.

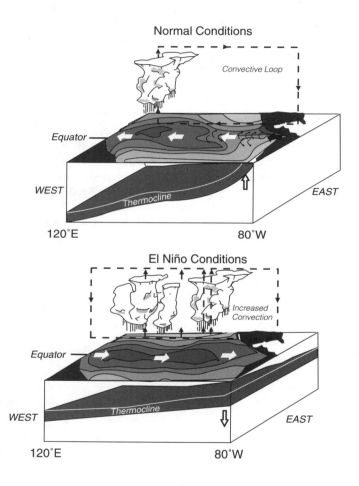

Figure 1. Schematic of ENSO. Contours indicate isotherms, with the increasingly darker fields near the equator showing increases in temperature. The top panel illustrates normal conditions in the tropical Pacific Ocean, while the lower one illustrates the warm phase of ENSO. (Courtesy of M. McPhaden, NOAA/PMEL.)

INTRODUCTION

EL NIÑO AND THE SOUTHERN OSCILLATION (ENSO)

In the western tropical Pacific, the sea surface is always warm (around 29°C), the sea-level pressure is low, and the precipitation is heavy (see Figure 1). In the eastern Pacific, some ten to fifteen thousand kilometers to the east, the situation is very different. There the water is normally cool (21°C to 26°C), the sea-level pressure is high, and the precipitation is low. The cold water in the region of the equatorial eastern Pacific (the "cold tongue") persists throughout the year but is warmest in April. This normal gradient in sea surface temperature along the equator is associated with westward winds. Although seasonal variations of this pattern occur, conditions are usually warm and wet in the west, and cool and dry in the east.

Occasionally, however, the warm pool in the western tropical Pacific begins to spread eastward. Accompanying these changes in sea surface temperature, the regions of low sea-level pressure and heavy rainfall move eastward with the warm pool, and the eastern and central Pacific become warm and rainy while the western Pacific becomes somewhat cooler and drier. The earliest recognition of these changes was an awareness of unusually warm water appearing from the north, off the coast of Peru, as an enhancement of the normal

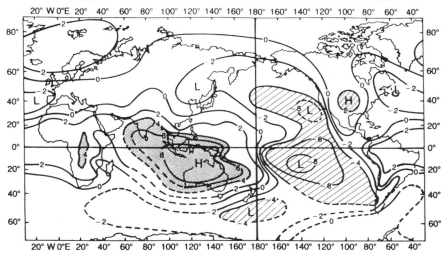

Figure 2. Correlations of annual-mean sea-level pressure with the pressure at Darwin, Australia. (From Trenberth and Shea 1987, reprinted by permission of the American Meteorological Society.)

Christmas warming. This warming of the coastal waters off Ecuador and Peru became known as El Niño. Later, it was recognized that the warming off Peru is part of a far more extensive warming, which extends across the entire tropical Pacific.

The Southern Oscillation is a widespread positive correlation of sea-level pressure anomalies centered near northern Australia that surround the warm pool in the western Pacific and extends considerably into the Indian Ocean, with anomalies of opposite sign centered east of Tahiti in the eastern and central Pacific (see Figure 2). These aspects of the Southern Oscillation have been known for more than 70 years (Walker 1924). However, not until the work of Bjerknes in the 1960s was it recognized that the expansion or contraction of the warm pool and the decreased or increased pressure in the central Pacific could both be part of a common phenomenon, now called ENSO (El Niño and the Southern Oscillation). A complete review has been provided by Philander (1990).

The term El Niño is still associated with the warm phase of ENSO. The cold phase is referred to sometimes as La Niña. Although these terms are common in the literature, the preferred designations are warm phase and cold phase of ENSO, referring to the temperature anomalies of tropical Pacific sea surface temperature; anomalies in other places may have opposite sign and different timing.

CONCEPT OF THE *TOGA* PROGRAM

Even the seemingly trivial prediction that conditions for each season or month will be close to that period's climatological mean has great social and economic value. All of us make plans based on the expectation of the normal progression of the seasons. Especially around the tropical Pacific, however, some years are significantly different from normal, confounding reasonable expectations and making planning for the near future especially difficult. Before the TOGA Program formally began at the start of 1985, the tropical Pacific and its overlying atmosphere were so poorly observed that it was impossible to detect even the largest warm phase of ENSO we have seen so far—in 1982–83—until it was well underway.

Although several programs conducted in the 1960s and 1970s gathered data on various atmospheric and oceanic aspects of the tropical Pacific Ocean, not until the early 1980s did the importance of ENSO became more generally appreciated. Research indicated correlations between the Southern Oscillation and the climate of the middle latitudes, thereby raising hopes that predictions of variations in the Southern Oscillation, and more generally of ENSO, would lead to useful predictions of middle-latitude climate. It was from this hope that the idea for the TOGA Program arose. When the TOGA Program formally closed at the end of 1994, we could obtain, using a computer and a telephone, information on the state of the surface and subsurface tropical Pacific for the prior day. As a result of the TOGA Program, not only can we access observations of current conditions in the tropical Pacific, but we can also skillfully predict aspects of the evolution of those conditions a year in advance.

PURPOSE OF THIS REPORT

The purpose of this report is to take a careful retrospective look at the TOGA Program from a U.S. perspective, to examine its elements and organization, to see what it has accomplished and where it has fallen short, to document the legacies it has left, and to present the opportunities it has created for programs to follow. The progress of the TOGA Program within the United States has been amply documented, from the original scientific plan (NRC 1983), through its implementation strategy (NRC 1986), to a mid-life review (NRC 1990), with an examination of the role of TOGA within the World Climate Research Programme (NRC 1992), and a review of the TOGA Observing System (NRC 1994a). This report joins its predecessors in the process of documenting and reviewing the TOGA Program. A set of scientific review papers on TOGA will be published in 1997 as a special issue of the *Journal of Geophysical Research*.

Different audiences will look to this document for different reasons. Those seeking a quick overview might read just the Summary and the bold-faced synopses at the beginning of each chapter and the beginnings of the sections of chapter 3. Policy makers will want to read the Summary, chapter 1, chapter 2, the last section of chapter 4, chapter 6, and chapter 7. Those interested in technical issues regarding ENSO should concentrate on chapters 2, 3, and 4. Researchers and program managers seeking to guide future work on short-term climate variations would profit from reading the entire report. The TOGA Panel intends that graduate students, among others, not only be able to read this report, but also that they be able to develop a picture of the breadth of the science and organization associated with a "big science" effort that was deeply connected to international programs. This intention has required judgments about how much material to include, in addition to the particulars of the U.S. TOGA Program. While this document is fundamentally about the U.S. contribution to the TOGA Program, completeness and continuity required that a coherent story be told, even though more has been included than can be attributed to the U.S. efforts designated for TOGA.

While the orientation of this document is fundamentally retrospective, it is designed to have an impact on future programs currently being planned to extend TOGA. The advisory structure provided by the NRC, in the form of the

Advisory Panel for the TOGA Program, was an important element in developing the program. The panel helped hold to the program together, raised the visibility of the program when TOGA had to contend with ups and downs of funding, and served as a community representative for dealing with the agencies and with the large community of TOGA scientists. The management structure of TOGA evolved into a set of national and international entities that cared for the funding and performance of the program and coped with the wide variety of obstacles that inevitably arise in a program of this size and complexity. And finally, the community of scientists provided the inexhaustible creativity and enthusiasm that carries any program through those periods when progress seems impossible.

The end of the ten-year TOGA Program is a milestone in global climate research. It is the first major element of the World Climate Research Programme to reach its conclusion, and marks significant progress toward a more interdisciplinary view of the global climate system. TOGA accomplished much. The program oversaw the building of an observing system for the tropical Pacific Ocean. It sponsored a series of process experiments, culminating in the massive TOGA COARE (Coupled Ocean–Atmosphere Response Experiment) field program. TOGA participants demonstrated the predictability of seasonal-to-interannual climate variations, developed prediction systems to exploit this predictability, and used these systems to predict climate variations in and over the tropical Pacific Ocean. The infrastructure and management tools of TOGA combined achievements in observations, modeling, and process experiments into a whole greater than the sum of the parts. Indeed, part of the success of TOGA was the forging of links between the various modeling and observational activities, and recognizing their mutual interdependence. The TOGA experience can therefore be viewed as a guide for existing programs and a prototype for the future.

For any climate program of finite duration, the challenge is to extract from the program what is most useful and enduring, and to discard what has proven unworkable. The useful parts should then be transformed and institutionalized into permanent observing, modeling, and prediction efforts. This report will aid in addressing this challenge. It will review the scientific basis upon which the TOGA Program was formed and will describe the progress in conducting process studies, in building an observing system, in fostering coupled atmosphere–ocean modeling, and in developing theory and techniques for making short-term (i.e., a season up to a couple of years in advance) climate predictions. This report will also examine the human, technological, and organizational means by which TOGA's progress has occurred. The newly opened possibilities for the application of climate prediction will be described as they relate to ENSO prediction. The scientific accomplishments and the changes in the communities of scientists brought about by TOGA will also be discussed. Finally, this document will indicate what TOGA did not accomplish. Understanding of these unfinished tasks will point to questions and problems for the future.

2. Growth of the TOGA Program

In the 1960s, Jacob Bjerknes was the first to link El Niño and the Southern Oscillation. He also suggested that the combined ENSO phenomenon resulted from dynamics that coupled the atmosphere and the upper ocean in the region of the equatorial Pacific. Coherent research efforts on ENSO developed during the 1970s, motivated in part by recognition of disruptive effects in the Americas from climate variations thought to be associated with ENSO. The intense ENSO warm event of 1982–1983 served as an impetus for the organized international program that became TOGA.

The TOGA Program grew out of studies of interactions between the atmosphere and ocean, especially the work of Jacob Bjerknes. Early observational, theoretical and modeling efforts to understand the dynamics of ENSO brought particular attention to the tropical Pacific region. Because of the large interannual variations in climate that seemed to result from ENSO, many scientists were willing to commit themselves to the planning process and the organized effort that became TOGA.

The scientific developments that ultimately led to the establishment of the TOGA Program and the highlights of early program plans and documents for TOGA have been presented previously (NRC 1990). Here we highlight the development of the scientific concepts and the events that shaped the plan for U.S. participation in the TOGA program, as it began in 1985.

ENSO: A Coupled Ocean–Atmosphere Phenomenon

The origin of the empirical and prediction studies of the TOGA program can be traced to the work of Sir Gilbert T. Walker, who assumed the post of Director General of the Observatory in India in 1904. In a region where monsoon failure and famine were disastrously coupled, Walker set about to understand and improve forecasts of the interannual variations of the Indian monsoon. Over the next three decades, Walker began to calculate correlations between the time series available to him—sea-level pressure, rain amounts, air temperature, sunspot activity, and others. In the process of performing these empirical studies, Walker established the existence of the Southern Oscillation as a global spatial pattern of interannual climate variations with identifiable centers of action. He was able to eliminate solar variability as a major contributor to the oscillation. Walker strongly suspected that oceanic processes were responsible for the oscillation, but was unable explore his ideas with the available data and thus missed the connection to El Niño (Walker 1924, Walker and Bliss 1932).

The first major breakthrough in understanding the mechanism of Walker's Southern Oscillation was to come forty years later. In the 1960s, Jacob Bjerknes began an examination of the meteorological conditions associated with El Niño, an interannually varying intrusion of warm equatorial waters along the western coast of South America. The warm intrusions had major economic impacts on the tuna fishing industry in the affected region. Bjerknes (1966) quickly recognized that El Niño was connected with large-scale fluctuations in trade-wind circulations in both the northern and southern hemispheres of the Pacific sector. He soon connected these trade-wind fluctuations to the Southern Oscillation (Bjerknes 1969).

Bjerknes's most important contribution was the reasoning he used in explaining the coupling between the oceanic and atmospheric circulations. On the basis of empirical evidence, Bjerknes hypothesized that El Niño and the Southern Oscillation are the result of the coupling between the east–west atmospheric circulation in the Pacific sector and also a coupling between the current and thermal structure of the upper ocean in the eastern equatorial Pacific Ocean. He observed that when the trade winds are strong (the "normal" condition), relatively cool equatorial water extends from the South American coast to the central Pacific. Bjerknes attributed the cool waters to equatorial upwelling caused by easterly wind stress acting on the ocean surface. He reasoned that the resulting pattern of sea surface temperature reinforces the strength of the trade winds by favoring large-scale atmospheric cooling, descent, and cloud-free conditions over the equatorial eastern Pacific, accompanied by large-scale ascent with relatively large amounts of precipitation, convective clouds, and atmospheric heating over the central and western equatorial Pacific. In the equatorial region, the east–west atmospheric heating differences would be expected to drive an east–west atmospheric overturning, which Bjerknes named the Walker circulation.

Bjerknes was able to link major decreases in the strength of the east–west gradient in equatorial sea surface temperature to decreases in the strength of the Walker circulation. He also linked changes in the gradient of sea surface temperature to disturbances in the planetary-scale atmospheric wave pattern over the sector covering the North Pacific Ocean and North America. This new information justified the cautious use of equatorial oceanic and atmospheric conditions for experimental climate forecasts (Bjerknes 1966, 1969). Bjerknes concluded that the variations in atmospheric heat input from the equatorial ocean were responsible for the observed interannual atmospheric fluctuations. The formulation of this hypothesis marked the beginning of a new era of climate studies in which the tropical ocean and global atmosphere were analyzed and modeled as coupled components of the global climate system.

EMERGENCE OF A COHERENT EFFORT (1970–1984)

The studies of Walker and Bjerknes established the coupled nature of El Niño in the ocean and the Southern Oscillation in the atmosphere as a single phenomenon now called ENSO. However, there was a need for an observing system and quantitative models that were capable of describing their interaction. The Bjerknes hypothesis provided a conceptual model of the feedbacks between the ocean and atmosphere that maintained the extremes of ENSO. The hypothesis emphasized the importance of sea surface temperature in producing the observed atmospheric anomalies during ENSO. However, many gaps in understanding remained, such as the mechanisms that control the initiation, development, termination, and irregular occurrence of warm events. The testable physical hypothesis of Bjerknes, however, provided a framework for the development of long-term observational strategies, process studies, and modeling research that would all join empirical studies to become major elements of the TOGA program (see Figure 2 in NRC 1990).

Nature provided scientists and policy makers with ample justification for continued scientific efforts. The 1972 and 1976 warm anomalies of El Niño along the South American coast were accompanied by alarming declines in the anchovy population. The economic repercussions of the anchovy decline were felt in the global economy (Barber 1988). The 1976–77 northern winter, which coincided with the latter warm episode, brought drought to California and record cold, accompanied by fuel shortages, to much of the central and eastern United States (Canby 1984). This coincidence of events raised public consciousness about possible connections, although some of these events were probably not related to El Niño. A 30-day-hindcast simulation (Miyakoda *et al.* 1983) of a blocking event during that winter helped make the case for the possibility of long-range dynamical forecasts.

A major breakthrough in the understanding of ENSO was the development of a more realistic view of the role of ocean dynamics in the evolution of upper-ocean thermal structure. Observational programs played a key role. During the 1970s, scientists involved in the North Pacific Experiment (NORPAX) Program (1971–1980), funded by the National Science Foundation and the U.S. Navy's Office of Naval Research, established the first oceanic monitoring system based on expendable bathythermograph (XBT) instruments deployed by ships of opportunity and island-based sea-level monitoring. Six major oceanic field programs begun during the pre-TOGA period—INDEX (Indian Ocean Experiment, 1976–1979), Hawaii to Tahiti Shuttle Experiment (1979–1980), EPOCS (Equatorial Pacific Ocean Climate Studies, 1979–1994), PEQUOD (Pacific Equatorial Ocean Dynamics Experiment, 1981–1983), SEQUAL (Seasonal Response of the Equatorial Atlantic, 1983–1984), and Tropic Heat (1984–1987)—provided an observational basis for a better understanding of the annual cycle and interannual variability of the three tropical oceans. These programs focused attention on the importance of equatorial ocean dynamics for ENSO.

Wyrtki (1975) supplied a key piece of the ENSO puzzle. He recognized the importance of the dynamical response of the upper equatorial Pacific Ocean to the large-scale weakening of the trade winds in the central Pacific during the onset of a warm event. On the basis of empirical evidence, Wyrtki hypothesized that the warm equatorial surface waters that developed at the western side of the basin during the period of abnormally strong trade winds would "surge" eastward, probably in the form of an equatorial Kelvin wave[*], depressing the thermocline as it passed. This would cause an abrupt rise in sea level and an increase in sea surface temperature through a reduction in the effects of upwelling on the eastern side of the basin. The Kelvin wave generated by this surge would cross the Pacific in approximately 50 days, presenting a potentially predictable dynamical feature of ENSO with a seasonal time scale. On the basis of this hypothesis linking temperature and wind anomalies, Wyrtki et al. (1976) made the first ENSO forecast. A series of cruises (the "El Niño Watch") was conducted in 1975 off the coast of Ecuador and Peru, but a warm event did not occur.

The reality of equatorially trapped Kelvin waves in the ocean was the subject of intense debate among theoreticians and observers, but was firmly established by analysis of sea-level and subsurface data from NORPAX, EPOCS, and PEQUOD. Evidence was found that connected the waves to forcing by wind stress (Knox and Halpern 1982, Eriksen 1982, Eriksen et al. 1983, Lukas et al. 1984, Mangum and Hayes 1984). This finding provided justification for interpreting the low-frequency variability in terms of long equatorial waves with low-order baroclinic structures having maximum amplitude in the upper ocean (see, e.g., Eriksen 1985). In contrast to features in the middle- and high-latitude oceans, large-scale features of the equatorial ocean circulation appeared to be linked to changes in the surface wind stress on monthly-to-seasonal time scales.

---

[*] Equatorial Kelvin waves are disturbances of thermal structure and currents, usually large-scale along the equator, trapped within a few degrees latitude of the equator. They propagate eastward at roughly 1–3 m/s. A different class of tropical disturbances, known as equatorial Rossby waves, propagates westward even more slowly. The region where these types of Kelvin and Rossby waves can exist, roughly within ten degrees latitude of the equator, is known as the "equatorial waveguide". See Gill 1982 for a more detailed explanation.

Simplified dynamical models of the equatorial upper ocean were found to reproduce the most important observed features of interannual sea-level and thermocline-depth variability, when the model ocean was driven by either idealized or observed wind-stress patterns (Hurlburt et al. 1976, McCreary 1976, Busalacchi and O'Brien 1981, Cane 1984). Not only did the simulated low-frequency oceanic waves generated by the imposed wind-stress anomalies affect the local region of the equatorial ocean where the waves were generated, but the waves could propagate systematically along the equatorial wave guide to remote coastlines.

Ocean modeling studies also provided a better understanding of the large-scale adjustment of an equatorial ocean basin to time-varying wind forcing. Philander and Pacanowski (1980) demonstrated that the adjustment time scale of the equatorial upper ocean to uniform zonal wind forcing depends on the longitudinal width of the equatorial channel, and, furthermore, that low-order baroclinic Kelvin and Rossby waves may play a dominant role in the adjustment process. The dynamical processes by which equatorial waves are reflected at boundaries on the eastern and western sides of the equatorial oceans were investigated (Cane and Moore 1981, Anderson and Rowlands 1976, Cane and Sarachik 1977, Moore and Philander 1977, Clarke 1983, Cane and Gent 1984). A simulation of the response of a bounded equatorial ocean to a uniform meridional wind forcing demonstrated that cross-equatorial flow over the eastern equatorial Pacific Ocean might also play an important role in determining the structure and variability of the "cold tongue" of sea surface temperature extending off the South American coastline (Philander and Pacanowski 1981b).

In parallel, atmospheric scientists were coming to a clearer understanding of the nature of thermally forced circulations in the tropics. Gill (1980) pointed out that thermally forced equatorial-wave solutions discussed by Matsuno (1966) and Webster (1972) bore a striking resemblance to the gross horizontal structure of climatological wind fields in the vicinity of large-scale tropical atmospheric heat sources. Zebiak (1982) demonstrated that when it is assumed that anomalous atmospheric heating is directly coupled to sea surface temperature anomalies, forced equatorial waves in the atmosphere can explain a significant fraction of the observed wind anomalies over the central and western Pacific Oceans during the different phases of ENSO.

The coupled nature of ENSO was beginning to emerge. It was clear that observed anomalies of tropical sea surface temperature were able to generate plausible anomalies of tropical atmospheric circulation, and that anomalies of atmospheric surface wind stress were able to generate plausible fields of anomalies of both equatorial upper-ocean circulation and sea surface temperature. These provided the elements of a feedback loop between the ocean and atmosphere. A linear stability analysis by Philander et al. (1984) demonstrated that a simplified model of the tropical ocean and atmosphere system might contain large-scale propagating—possibly unstable—modes not present in either the ocean or the atmosphere alone. The working hypothesis for ENSO was thus extended to include not just oceanic Kelvin and Rossby waves, but large-scale Kelvin-like or Rossby-like modes of the tropical coupled ocean–atmosphere system.

During the pre-TOGA period, atmospheric scientists also began intensive use of both global data sets and more complex models. They quantitatively examined the effects of regional equatorial circulation and heating anomalies on the atmospheric circulation in remote regions, "teleconnections" in the terminology of Bjerknes. Applying empirical techniques to the study of global atmospheric data sets, investigators succeeded in isolating patterns of the middle- and upper-tropospheric planetary waves associated with ENSO (Horel and Wallace 1981, Trenberth and Paolino 1981, van Loon and Madden 1981). Building on the early work of Rowntree (1972, 1976), modelers attempted to reproduce these patterns with the use of complex atmospheric general-circulation models. In these numerical experiments, "control" integrations were performed with a prescribed sea surface temperature to obtain the normal time-averaged state of the model. The general-circulation models were then perturbed by introducing anomalous patterns of sea surface temperature as lower boundary conditions.

Statistically significant correlation patterns were obtained using several general-circulation models (Julian and Chervin 1978, Keshavamurty 1982, Blackmon *et al*. 1983, Shukla and Wallace 1983, Nihoul 1985—especially within the last, Cubash 1985, Boer 1985, von Storch and Kruse 1985, Palmer 1985, and Fennessy *et al.* 1985). Although a definitive explanation for the mechanism causing anomalous patterns of atmospheric planetary waves was lacking, theoretical studies conducted during this period indicated that the energy of planetary-scale waves generated by a regional process in the equatorial zone was able to propagate to remote regions of the global atmosphere within a few days under certain conditions. The striking resemblance between the observed and theoretically obtained wave patterns led to a working hypothesis that Rossby-wave propagation might be responsible for the "teleconnections" between the tropical Pacific and other locations (Hoskins and Karoly 1981; Webster 1981, 1982; Simmons 1982; Simmons *et al.* 1983). Results of atmospheric and oceanic modeling efforts, as well as statistical analyses of the lag-lead relationships between sea surface temperature and climate anomalies over North America (Barnett 1981), indicated that deterministic ocean models driven by observed wind stress might be used to produce useful forecasts a season in advance, even if it were not possible to predict the evolution of an entire ENSO cycle.

DEVELOPMENT OF THE TOGA PROGRAM

In 1982, the international climate research community established a study group under the leadership of the late Adrian Gill to examine short-term climate variability. The group considered the growing evidence that a significant part of climate variability on seasonal-to-interannual time scales could be understood in terms of interactions between the three major tropical oceans and the global atmosphere. ENSO was clearly the largest and most coherent signal in the seasonal-to-interannual range. Other examples of interest to the international community included the connection of anomalies of sea surface temperature in the tropical Atlantic with precipitation over northeast Brazil and the Sahel (Hastenrath and Heller 1977, Moura and Shukla 1981), and the connections of anomalies of sea surface temperature in the eastern Indian Ocean (Streten 1983) with rainfall anomalies over Australia. It was recognized that a longer-term investigation would be needed to verify the working hypotheses. The study group also recognized that such an investigation might be feasible because the seasonal-to-interannual oceanic variability appeared to be confined primarily to the upper ocean of the tropics, while the extratropical ocean circulation on these time scales appeared to be relatively insensitive to remote atmospheric influences. This recognition narrowed the scope of the investigation and provided a scientific basis for setting priorities on the observational and modeling activities.

When the international TOGA Scientific Steering Group was formed in 1983 to define the TOGA program's scientific and technical scope, the NRC's Climate Research Committee was already engaged in drafting its own scientific plan for ENSO research (NRC 1983). By this time, ENSO research had emerged as a major theme of the U.S. National Climate Program. It is interesting to note that ENSO research had not been emphasized in the planning documents that supported the drafting and passage of the National Climate Program Act in 1978. ENSO research emerged from the grass-roots effort of a relatively small number of scientists, supported by traditional funding sources. These funding sources included grant sections of the National Science Foundation and mission-oriented agencies such as the National Oceanic and Atmospheric Administration, which has responsibility for monitoring and predicting daily-to-seasonal weather and climate. Given the interests of the U.S. scientific and policy-making communities, it was agreed from the outset that the primary U.S. contribution to the international TOGA field effort would be made in the Pacific and would focus on ENSO. The time scale of phenomena examined under the program would be seasonal to interannual, and the measure of progress in understanding would be the extent to which global coupled ocean–atmosphere circulation models, when initialized by state-of-the-art observations, would be able to model and predict oceanic and atmospheric phenomena.

The 1982–83 El Niño exerted a very significant influence on the scientific planning for TOGA. In addition to focusing attention on the Pacific Ocean, this event surprised the scientific community with its magnitude and manner of evolution. However, the very existence of the large warming in the Pacific was obscured by the 1982 eruption of El Chichon. The aerosol veil resulting from the eruption obscured viewing by the AVHRR (Advanced Very-High Resolution Radiometer) satellite instrument and made satellite-derived estimates of sea surface temperature unreliable. A composite of six warm events after 1949 had shown a tendency for warm events to begin, peak, and end at preferred times of year, as well as to exhibit the hypothesized connection between warm El Niño events and the Southern Oscillation (Rasmusson and Carpenter 1982). The 1982–83 event didn't follow this pattern. It skipped the typical "onset phase" characterized by anomalously warm (El Niño) conditions along the South American coast during April, which typically spread westward along the equator toward the central Pacific (Rasmusson and Wallace 1983). Apparently, it was possible for individual ENSO cycles to differ significantly in detail from the composite picture. The concept of a rapid-response observational strategy (see NRC 1983) was therefore abandoned in favor of continuous, "real time" monitoring of the equatorial Pacific.

SCIENTIFIC PLAN FOR TOGA (1985)

The challenge for the organizers of the TOGA Program was to formulate a scientific plan that would reflect a new way of thinking about the climate system. It would not be possible to make fundamental progress by consideration of either the atmosphere or the oceans alone. Observational systems and models were needed to describe the system encompassing the tropical oceans and the global atmosphere, a system in which the processes determining sea surface temperature, atmospheric heating, surface wind stress, and ocean circulation were related to each other in feedback loops (see Figure 1).

The scientific objectives adopted in the international plan (WCRP 1985) were:

1. To gain a description of the tropical oceans and the global atmosphere as a time-dependent system, in order to determine the extent to which this system is predictable on time scales of months to years, and to understand the mechanisms and processes underlying that predictability;
2. To study the feasibility of modeling the coupled ocean–atmosphere system for the purpose of predicting its variations on time scales of months to years; [and]

3. To provide scientific background for designing an observing and data transmission system for operational prediction if this capability is demonstrated by the coupled ocean–atmosphere system.

While the formal objectives of the international TOGA Program did not mention ENSO specifically, it had been known that the ENSO phenomenon dominated the interannual variability of sea surface temperatures in the tropical Pacific (Weare *et al.* 1976). Furthermore, it was believed, although not demonstrated until recently (e.g., Kawamura 1994, Mann and Park 1994), that ENSO drives the dominant seasonal-to-interannual signal in sea surface temperature globally. In fact, the 1983 NRC report that presented a science plan on short-term climate variations strongly emphasized ENSO, even in the report title. The early decision, therefore, was to concentrate on the big seasonal-to-interannual signal— ENSO in the tropical Pacific. It was also known that ENSO occurs irregularly with a spectral peak at about a 40-month period (Rasmusson and Carpenter 1982). The irregular nature of ENSO argued for an open-ended program that would continue until a few ENSO cycles could be observed in detail. Practical considerations of national and international program management dictated that the program have a predetermined completion date. It was decided that TOGA would have a ten-year duration, beginning on 1 January 1985. This was perhaps the shortest period that would permit significant progress by examining more than one ENSO cycle and the longest period likely to be approved by the sponsoring U.S. agencies and other participating governments, even if it was not sufficient to capture the full range of ENSO behavior.

The scientific objectives clearly established the performance of coupled ocean–atmosphere models as the ultimate measure of success for the TOGA Program. Consistent with the international objectives, the scientific plan for U.S. participation in TOGA (NRC 1986) identified improvement of atmospheric and oceanic circulation models, and the successful coupling of these components, as central elements. It was also recognized that model development and predictability studies would be impossible without a major improvement in the observing system over the data-sparse areas of the tropical oceans. Prediction schemes would require a specification of the initial thermal and current structure of the upper ocean. The fragmentary oceanic and atmospheric observations during the pre-TOGA period had left unanswered many questions about the structure and dynamics of the seasonal-to-interannual variability of the coupled system. An observational system of unprecedented spatial and temporal coverage would be required to guide model development, validate model performance, and permit the empirical description of at least one ENSO cycle. The U.S. contribution to the TOGA observing system would focus on the Pacific Ocean, and on the ENSO phenomenon in particular.

Finally, it was envisioned that one or more process studies would be required for significant improvement of model parameterizations of key physical processes, such as upper-ocean mixing, atmospheric turbulence and convection, and interactions between clouds and radiation. Process studies would also permit sorting out of the relative importance of the myriad atmospheric and oceanic phenomena with similar or shorter time scales that might play a role in altering the behavior of coupled modes of seasonal-to-interannual variation. The drafters of the original science plan (NRC 1983) could not determine the optimal location for the process experiment(s)—eastern, central, or western Pacific—and therefore recommended a separate experiment for each. As it turned out, an experiment was performed in each of these regions. The major process study during TOGA, COARE, was performed in the western Pacific, the region thought to be most important for the development of ENSO.

3. COMPONENTS OF THE U.S. TOGA PROGRAM

The U.S. TOGA Program examined seasonal-to-interannual climate variability using observational, theoretical, and numerical techniques. In this chapter, we report on the major observing, process-study, and modeling and prediction efforts of TOGA. TOGA organized a large observational program, many of the instruments of which remain in place as the TOGA Observing System. Process studies, especially TOGA COARE, were performed to fill gaps in our knowledge of crucial dynamics on smaller spatial and temporal scales. Models, especially coupled models, and prediction systems were developed to address many aspects of ENSO. The interrelations are summarized by Figure 3, which indicates that the components all interact to motivate and justify each other. These components are now beginning to interact with applications of climate information and studies of the impacts of climate variations.

Over the ten-year lifetime of TOGA, U.S. federal agencies spent over $230 million to fund U.S. efforts directly contributing to the international TOGA Program. This funding represented roughly half of the direct contributions from the eighteen nations that were members of the Intergovernmental TOGA Board

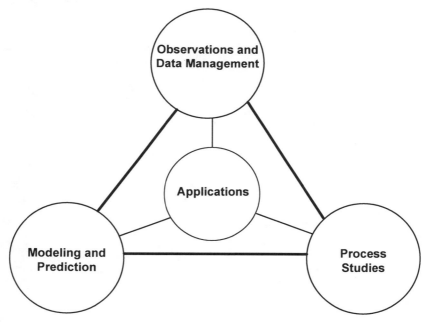

Figure 3. The major conceptual components of TOGA.

(see p. 121). A partial enumeration of international funding for TOGA can be found in IGFA 1993. Figure 4 provides a bar chart of U.S. spending by year, which ranged from $15 million to $39 million. For comparison, the 1994 budget for the USGCRP was $1443 million (Subcommittee on Global Change Research 1995), of which $37.5 million was directly attributed to TOGA. The near-doubling of the budget from 1989 to 1992 reflects the buildup for the TOGA COARE field program (see p. 51) of 1992–93. These totals do not include the costs of aircraft time and ship time, which were budgeted separately.

The TOGA Program benefited from activities related to predicting seasonal-to-interannual climate variations that were not considered part of

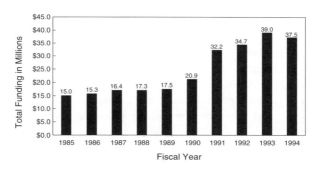

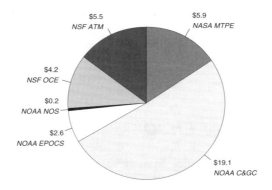

FY 1994 ($ in Millions)

Figure 4. Funding for U.S. TOGA. The bar chart shows total funding, by year, for U.S. efforts contributing to the TOGA Program. The pie chart shows U.S. funding, by program, for the last year of TOGA. See the text for descriptions of activities that were included or excluded from the totals, and for the program abbreviations. (Data courtesy of M. Patterson, NOAA.)

TOGA. For example, no costs of satellite-based instruments are included in the totals given above and in the figure. However, TOGA made use of operational meteorological satellites operated by NOAA (National Oceanic and Atmospheric Administration) and research satellites operated by NASA (National Aeronautics and Space Administration) for multiple scientific purposes. The totals for TOGA do not include most of the international meteorological observing system on which TOGA relied, but do include the specialized observing system that was deployed as part of TOGA in the tropical Pacific. The TOGA Program also benefited from work by the National Weather Service to provide operational long-term outlooks and to develop numerical models for short-term climate variations, but those costs are not included.

Several U.S. federal agencies participated in TOGA. A breakdown of their contributions for the final year of the program is shown in the Figure 4 pie chart. A similar chart for 1989 can be found in NRC 1990, which also includes a discussion of the funding mechanisms for TOGA. The statistics on which these figures were based came from an annual survey conducted by the U.S. TOGA Project Office. Funding levels include only resources requested and directed specifically for TOGA research. NOAA funding came from three offices: the NOAA Climate and Global Change Program (labeled C&GC, which includes its predecessor, the NOAA TOGA Program), the NOAA Equatorial Pacific Ocean Climate Studies (EPOCS) Program, and the National Ocean Service (NOS). National Science Foundation (NSF) funding came from the Ocean Sciences Division (OCE) and the Atmospheric Sciences Division (ATM). NASA funding came from the Mission to Planet Earth (MTPE) Program. The Office of Naval Research participated only during 1992–1993, for TOGA COARE, and is therefore not shown in the pie chart for the last year of the program. Participation by various agencies in specific field programs is noted below in this chapter.

OBSERVATIONS OF ENSO

> The TOGA Observing System was initially designed to observe the evolving warm and cold phases of ENSO, and then provide the resulting data to scientists immediately. As predictions of the phases of ENSO began to show skill, priorities shifted to measuring those quantities most useful for initializing and evaluating coupled models. We now have the ability to measure the quantities of highest priority—sea surface temperature, surface winds, and subsurface thermal structure—plus some other quantities of interest.

If we were to stand on the equator with our backs to the coast of Ecuador and begin moving westward along the equator at walking speed—coincidentally, about the speed of the fastest ocean Rossby mode—it would take about nine months to reach the east coast of Halmahera Island in the far western Pacific. Over this vast expanse of ocean, only a few Volunteer Observing Ship (VOS) lines were in place throughout the 1960s and 1970s (see Figure 1 of Rasmusson and Carpenter 1982). VOS lines are series of observations of temperature, humidity, winds, cloudiness, and upper-ocean conditions made on a voluntary basis by the crews of non-research ships while the ships are traversing their normal routes. These limited observing resources left vast regions of the equatorial Pacific, almost as wide as the entire equatorial Atlantic, unmeasured. The problem for observing ENSO was to measure enough of the tropical Pacific, both at the surface and at depth, and to do it frequently enough to resolve the evolving atmospheric winds and ocean thermal structure.

At the beginning of TOGA (see Figure 5, top), there was no obvious solution to this problem. It was envisioned in the original TOGA Plan (sections 2.1 and 2.2 of NRC 1983) that the VOS observations would be augmented by a small number of drifting thermistor chains and fixed moorings, by additions to the upper-air network of the World Weather Watch, and by satellite measurements. As explained below, the original concept of the TOGA Observing System proved inadequate, and more creative and extensive means had to be developed. During the TOGA years, additional systems were put into place to reach the goals of TOGA. By the middle of the program's life, the TOGA Observing System looked as shown in Figure 5, middle, and by the last half of 1994, the TOGA Observing System looked as shown in Figure 5, bottom.

In addition to the need to observe regions previously under-observed, there was also a need to arrange for the prompt transmission of observational data to scientists (transmission "in real time"). This requirement necessitated changes to oceanographic instruments, such as moorings that recorded data on tape but did not transmit them, as well as changes in the practices of oceanographers. The scientists involved in TOGA made enormous progress in the development of observing systems that provided data in real time.

TOGA Observing System

The TOGA Observing System is a primary legacy of the TOGA Program. The following discussion touches on the broad range of observations used to address the scientific objectives of TOGA. Some of these observations were not established by TOGA, but were used, championed, or sustained by scientists participating in TOGA. As noted above, the TOGA Observing System was initially designed to observe ENSO. As such, it concentrated on the tropical oceans,

TOGA in Situ Ocean Observing System
Pacific Basin

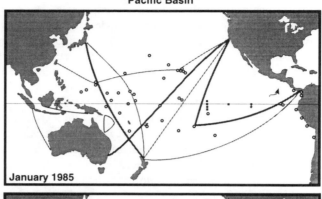

January 1985

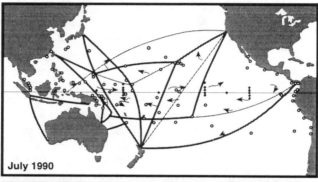

July 1990

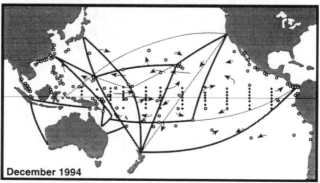

December 1994

Figure 5 (at left). The TOGA *In Situ* Pacific Basin Ocean Observing System. The top panel is as of the start of TOGA in January 1985, the middle panel as of 1990, and the bottom panel as of the end of TOGA in December 1994. Moored buoys with thermistor chains and wind-measuring equipment are shown with diamonds, and those with current meters added are shown with squares. Island and coastal tide gauges that reported to the TOGA Sea Level Center are shown with circles. Drifting buoys are indicated schematically with short arrows; each arrow represents ten actual drifters. Tracks of volunteer observing ships releasing expendable bathythermographs are indicated by long curved lines, thick lines representing 11 or more transects per year and thin lines representing 6–10 transects per year. By the end of TOGA, most measurements were being reported in real time via satellite data relays. (Courtesy of M. McPhaden, NOAA/PMEL.)

especially the Pacific. A review of the TOGA Observing System was also presented in NRC 1994a. Building on the sparse operational network in place at the beginning of the program, TOGA built a diverse system incorporating observations that were:

- meteorological and oceanographic,
- land-based and ocean-based,
- remotely sensed (mostly from space) and *in situ*,
- funded by operational agencies and by research programs,
- driven by predictive model requirements and by curiosity-driven desires for elucidation of physical mechanisms in the climate system,
- obtained by standard means and by novel technologies, and
- solutions to TOGA data needs and to the data needs of other scientific and operational constituencies.

The scope of the system was broader than the idealized (and not as yet established) observing system envisioned by TOGA Objective #3: "... an observing and data-transmission system for operational prediction...." Such a system, tuned to provide just the data required for "operational prediction"—not more, not less—remains an ideal. A method for optimizing the observing system, using clear results from accepted prediction models, is still in development. It is likely that this ideal system will include a number of substantial enhancements and augmentations of the present-day system, and perhaps some selective winnowing of observations that do not carry much information into the prediction process in proportion to their cost. A limit on any such winnowing is that while some particular measurement series may prove to be of marginal utility for TOGA purposes, the data time series may be of great value for some other set of scientific studies or operations, and those other constituencies should be included in discussions about the future modification of the observing system.

Table 1. TOGA data requirements

	Quantity	Spatial Resolution	Temporal Resolution	Accuracy
1	Upper-air winds	500 km (900 mb, 200 mb)	1 day	3 m/sec
2	Tropical wind profiles	2500 km (every 100 mb)	1 day	3 m/sec
3	Surface pressure	1200 km	1 day	1 mb
4	Total-column precipitable water	500 km	1 day	0.5 g/cm^2
5	Area-averaged total precipitation	2° latitude by 10° longitude	5 days	1 cm
6	Global sea surface temperature	2° latitude by 2° longitude	30 days	0.5 K
7	Tropical sea surface temperature	1° latitude by 1° longitude	15 days	0.3–0.5 K
8	Tropical surface wind (1)	2° latitude by 10° longitude	30 days	0.5 m/sec
9	Tropical surface-wind stress (1)	2° latitude by 10° longitude	30 days	0.01 Pa
10	Surface net radiation	2° latitude by 10° longitude	30 days	10 W/m^2
11	Surface humidity	2° latitude by 10° longitude	30 days	0.5 g/kg
12	Surface air temperature	2° latitude by 10° longitude	30 days	0.5 K
13	Tropical sea level	(2)	1 day	2 cm
14	Tropical ocean subsurface temperature and salinity	(3)	(3)	(3)
15	Tropical ocean surface salinity	2° latitude by 10° longitude	30 days	0.03 PSU
16	Tropical ocean surface circulation	2° latitude by 10° longitude	30 days	0.1 m/sec
17	Sub-surface equatorial currents	30° longitude (at 5 levels)	as available	0.1 m/sec

(1) While the accuracy values are given for 30-day averages, daily values are required for the resolution of 30–60 day oscillations.

(2) Resolution as permitted by available sites and satellite altimetry.

(3) See discussion in chapter 3 of NRC 1994a.

A good overview of TOGA requirements for observations and the methods that were used to meet some of those requirements can be found in sections 2–7 of the International TOGA Implementation Plan (fourth edition, International TOGA Project Office 1992). Table 1, extracted from that plan, shows the atmospheric and oceanographic quantities desired, with associated sampling intervals and accuracies in both space and time. The Implementation Plan also provides brief rationales for the numbers presented. Mostly, the accuracies and sampling requirements are based on observed amplitudes of ENSO signals in the ocean and atmosphere. Prediction models and observing-system simulation experiments (OSSEs) are not yet able to reliably guide the design of complete observing systems.

Sections 3–7 of the Implementation Plan summarize well the current status and near-term outlook for upper-air instruments, moorings, drifters, VOS, and island tide gauges, respectively. These were the principal tools of the TOGA Observing System for *in situ* measurements. Their condition at the end of 1994 is the TOGA legacy for observations insofar as TOGA played a role in creating or sustaining them. A particular measurement platform usually contributed to measurements of more than one quantity of interest. For example, the TOGA TAO moored array provides data on surface winds, surface-air temperature, sea surface temperature, surface atmospheric pressure, and upper-ocean thermal structure, with additional data on upper ocean currents at some of the array elements. Space-based observations were generally not designed specifically by or for the TOGA community, yet they have proven crucial for measuring certain basic quantities of the ocean and atmosphere, and will therefore be discussed as part of the system.

To explain how and why the TOGA Observing System developed as it did, we provide a history of the observing system with an emphasis on the overall motivation and on the scientific and technical problems that arose at crucial stages along the way. It begins with a discussion of how traditional types of oceanographic measurement platforms were developed to address the problems of TOGA. By the time TOGA ended, a variety of options were available to measure what were, by then, agreed-upon quantities of highest priority. A final section indicates how, and how well, these high-priority quantities were measured.

Volunteer Observing Ships

Merchant ships, recruited under the auspices of the World Weather Watch to take standard surface observations for meteorological purposes, have been the major source of surface data over the open oceans for many years, and undoubtedly will continue to be for some time. TOGA heightened our awareness of both the utility and drawbacks of volunteer observing ships (VOS) as a data source. It also advocated the continuation of VOS measurements and pointed out ways in which they could be improved and enhanced. Participants in TOGA championed, with some success, the identification of additional VOS candidates in data-poor tropical regions, improvement in the promptness and reliability of data communication using conventional high-frequency radio or newer links (e.g., communication satellites), the aggressive recovery of additional data for archiving from platforms such as fishing vessels that do not provide immediate data transmission, the upgrading of the accuracy of shipboard instruments and methods, and the development of hull-contact sensors for ocean measurements (thereby easing installation and calibration problems). Modernizing and bringing such a far-flung system to maximum effectiveness is not a one-shot process; continual pressure has been and will be required.

A particularly effective "add-on" to basic VOS surface observations has been the inclusion of regular expendable bathythermograph (XBT) observations of upper-ocean thermal structure to the routine of selected VOS that traverse scientifically critical routes. Much of this work has been planned and carried out in concert with the World Ocean Circulation Experiment (WOCE) because a given ship often crosses regions of significance to both programs. A joint TOGA and WOCE set of route charts and sampling-interval suggestions was produced. Not all the recommended routes were sampled, and even fewer were sampled as frequently as TOGA objectives required, but there has been progress in recent years. While the accumulated VOS reports are of great value in our knowledge of the background state and climatology of the ocean, the tracks are too wide apart in the tropics and visited too infrequently to permit the study of the evolving warm and cold states characteristic of ENSO (see the figure on p. 30 of NRC 1994a).

Surface Drifters

Lagrangian drifters closely follow the ocean currents in which they sit. They are an effective means of obtaining broad, basin-wide coverage of sea surface temperature, sea-level pressure, and near-surface currents. In the late 1970s, Doppler-ranging from the French navigational system ARGOS became operational on NOAA TIROS (Television Infrared Observation Satellite). It was a cost-effective technique for listening to and locating radio transmitters on the ocean surface. The development of TIROS spawned the design and construction of a large number of ocean-surface drifters for measuring ocean circulation and as platforms for a variety of meteorological sensors.

The drifters designed in the 1970s typically used large surface floats and "window-shade" drogues, a combination that degraded their ability to follow surface currents accurately and limited their useful lifetimes. During the planning of the World Climate Research Programme (WCRP 1983) several researchers realized that a redesigned drifter, one that was lightweight, long-lived, and readily deployable from ships, could make widespread measurements of sea surface temperature, surface pressure, and surface velocity to accuracies of ±0.3 K, ±1 mb, and ±1 cm/sec, respectively. By 1985, Draper Laboratory of the Massachusetts Institute of Technology, NOAA's Atlantic Oceanographic and Meteorological Laboratory (AOML) and the Scripps Institution of Oceanography (SIO) had produced competing drifter designs. During the period 1985–1989, field measurements of the water-following capability of the drifters were made using vector-measuring current meters (VMCMs) attached to the top and bottom of the drogue attached to the drifter (Niiler *et al*. 1987, Niiler *et al*. 1994). Several modeling studies of drifter behavior in steady upper-layer shear and linear wave fields were also carried out (Chereskin *et al*. 1989). These studies provided a basis for interpreting possible differences between the movement of the drifter and the movement of the water.

In 1978, under the auspices of the NOAA/EPOCS program, the deployment of small groups of drifters began in the tropical Pacific (Hansen and Paul 1984). Under the auspices of TOGA in 1988, a basin-wide process experiment, the Pan-Pacific Surface Current Study, began. Its technical objectives were, over a three-year period, to learn to use VOS to maintain a group of 160 drifters and to select the most robust elements from the competing drifter designs. Its scientific objectives were to obtain a basin-wide field of surface currents and sea surface temperature of the tropical Pacific for the purpose of studying processes that determine the evolution of sea surface temperature. By 1991, these objectives had been realized and a WOCE/TOGA Lagrangian Drifter design was published (Sybrandy and Niiler 1991). These new drifters had a half-life of over 400 days at sea. They were deployed routinely, with over 95 percent survival, from VOS by a single able-bodied seaman. Costs for tracking and communication were reduced to one-third of the costs incurred using the older design, with no reduction of data quality. New sensors were also developed. Barometers were added to the drifters in 1991. In 1992, new salinity sensors were deployed during TOGA/COARE.

At the end of 1994, there were nearly 500 drifters, deployed from VOS or aircraft, in the global ocean. The greater portion of these drifters reported data through the Global Telecommunication System (GTS) to operational meteorological and oceanographic centers. The data from these drifters were processed at the Drifter Data Center (housed at NOAA/AOML), and are available over the Internet using anonymous ftp (file-transfer protocol) with about a one-month delay. Fourteen countries contribute resources to this drifter program, and seven U.S.-government agencies use the WOCE/TOGA Lagrangian Drifter as an operational instrument.

The Global Drifter Program in support of climate research and monitoring is now in place. It is an element of both the Global Ocean Observing System (GOOS) and the Global Climate Observing System (GCOS). TOGA played the principal role in the inception and implementation of this drifter program. Of primary interest to TOGA were sea surface temperature data from drifters. These data are critical because they provide well-distributed benchmarks for adjusting and correcting the inaccuracies and biases in maps of sea surface temperature derived from other sensors (such as those on satellites or VOS). Of great interest in oceanography, but of somewhat less importance to ENSO predictions, is the ability of drifters to determine accurately, with minimal contamination from wind and other noise, large-scale patterns of the surface currents. Drifters are being constructed in and deployed by several nations, using both research and operational funding. A census of 250 units in the tropical Pacific was stated as a goal for TOGA, and at the end of 1994 was approximately met. An additional set of 50 meteorological drifters (compared with a stated requirement for TOGA of 100) in the southern oceans was sustained primarily to obtain surface meteorological data, especially sea level pressure, in this data-sparse region. These drifters, the so-called FGGE (pronounced "figgy", after the First GARP Global Experiment) buoys, have a different hull form, so their displacements are less useful than the new Lagrangian drifters for mapping surface currents.

Tide-Gauge Network

Tide gauges measuring sea-level changes have been maintained for many years in harbors and lagoons of Pacific islands. Some of these gauges date back to shortly after World War II. In 1972, NOAA shut down the upper-air and sea-level stations at Canton Island and Christmas Island, right at the beginning of the NORPAX program. This action was taken as the 1972–73 El Niño was beginning, creating an unfortunate break in critical time series that were more than 20 years long. Jacob Bjerknes, who had used the long time series from Canton Island in his research on ENSO, wrote a protest letter to NOAA in hopes that this key station would be maintained. His letter fell on deaf ears. Such was the lack of appreciation for climate monitoring at that time.

Klaus Wyrtki of the University of Hawaii, whose research using sea level made the case for the importance of ocean dynamics in climate (Wyrtki 1973), immediately went to work to rebuild the sea-level network of Pacific islands to ensure long time series for future climate research. He proposed a sea-level monitoring network for climate-research purposes. NSF funded the proposed network under the NORPAX program, and Wyrtki began installing tide gauges on various islands. In addition, Wyrtki expanded this network by forging relationships with numerous other Pacific nations that maintained tide gauges for their own purposes. The science that emerged from the sea-level measuring effort was an important part of the foundation upon which TOGA was built.

The newly established tide-gauge network monitored the large water-mass displacements during the 1976 and 1982–83 ENSO warm events, leading to the discovery of the very large horizontal scales of sea-level variability associated with these events (Wyrtki 1979, Firing *et al.* 1983). The original grant for Wyrtki's efforts was renewed for ten years under the International Decade for Ocean Exploration. Starting in 1982, support of the tide-gauge network was provided jointly by NSF, NASA, and NOAA under new funding for the upcoming TOGA program. In 1990, NOAA took over responsibility for funding the Pacific Sea Level Network. Expansion of the sea-level network into the Indian Ocean began in 1986. As of the end of 1994, 115 stations were transmitting data regularly; 79 were in the Pacific, 24 were in the Indian, and 12 were in the Atlantic. Of the total, 68 reported in real time. In the Pacific, almost all of the islands that might host gauges now have them; some islands present logistical problems that make it almost impossible to maintain a tide gauge. At the close of TOGA, the network was still being expanded in the Indian and Atlantic Oceans; the most recent addition was five tide gauges deployed by France in the tropical Atlantic. Satellite telemetry of the measurements has become easier; its further use can only enhance the value of the sea-level data for such purposes as near-real-time inclusion in operational ocean-circulation models.

The TOGA Sea Level Center (TSLC) was created in 1985. It was charged with collecting data from all data originators within the global tropics, processing and quality-controlling these data, and distributing them within eighteen months of collection. Originally, only daily-mean sea-level values were to be disseminated, but it soon became apparent that hourly data were required for adequate quality control and for resolving the physical quantities of interest. Also, monthly-mean products were originally planned and provided. The TSLC now makes a provisional sea-level data set from selected stations available with only a one-month delay. The rapid provision of *in situ* data has proven to be an important supporting contribution to the success of the U.S./France TOPEX (Ocean Topography Experiment)/Poseidon altimetric sea-level satellite mission. As of July 1994, the TSLC archives contained 3424 station-years of data from 289 sites, with half of the data added during the previous two years having come from the Indian and Atlantic Oceans. The TOGA sea-level data from tide gauges has become one of the most frequently requested data sets.

TAO Array

The establishment of the Tropical Atmosphere/Ocean (TAO) moored array of instruments in the tropical Pacific (see Figure 5, bottom) was one of the crowning achievements of the TOGA program. The array measures those quantities of highest priority for understanding and predicting ENSO. Its deployment was a great technical achievement. However, the basin-wide TAO array was not completed until very late in the TOGA program. It was conceived because all other existing means for real-time monitoring of the Pacific, whether *in situ* or remote, had proved inadequate to fulfilling the goals of TOGA.

Early in the planning for TOGA, it was recognized that the existing and planned XBT lines from the VOS program would not provide sufficient spatial and temporal coverage for determining the upper-ocean thermal structure as specified in the scientific plan for TOGA (NRC 1983). There were large gaps between some of the ship tracks, and temporal sampling in the equatorial region was inadequate. In order to address these problems, the earliest versions of the TOGA Implementation Plan called for the deployment of drifting thermistor chains, which would measure the temperature profile of the upper ocean. It was believed that the drag over the length of the chain would prevent rapid movement through the areas where the data were needed, and that chains could be easily launched from VOS. Efforts by NOAA's Pacific Marine Environmental Laboratory (PMEL) and SIO resulted in the deployment of several such drifters, but they did not work well. In particular, the commercial manufacturer chosen could not solve quality-control problems. (A French version of this device worked well in the Indian Ocean and during TOGA COARE.)

The ATLAS thermistor-chain mooring (described in detail below) was developed during the early 1980s, with support from the NOAA/EPOCS program, in parallel with the efforts to develop drifting thermistor chains. However, it was not until late in 1986, when the drifting thermistor chains were abandoned, that new plans were made to put a limited number of TAO moorings out along three longitudes. As the ATLAS mooring system demonstrated its usefulness, an ambitious plan was developed to expand the TAO array to cover the entire Pacific wave guide. This expanded array was expected to provide sufficient simultaneous wind and thermal-structure information to allow determination of the nature of the inadequacies in ocean models. (These inadequacies had been attributed to both poor specification of the wind forcing and to incorrect model physics.)

The development of the TAO array rested on earlier efforts for making unattended measurements in the open ocean. In the 1970s, NOAA/PMEL, the P.P. Shirshov Institute of Oceanography, and Woods Hole Oceanographic Institution all experimented with techniques for recording time series of upper-ocean current and temperature measurements using a surface buoy anchored to the bottom at depths of 4–5 km. All faced difficulties in deploying and maintaining surface moorings because of stresses on the mooring line caused by the shears of strong currents in the upper ocean. For example, in the Pacific, the eastward Equatorial Undercurrent, which has a maximum speed of 1 m/s at 100 m depth, flows beneath the westward South Equatorial Current, which has a maximum speed of 1m/s at the surface. The accuracy of wind measurements from a moored surface-following float (Halpern 1987a) and the accuracy of upper-ocean current measurements beneath such a float (Halpern 1987b) were acceptable for studies of the equatorial ocean.

The first collection of moored surface buoys to span the equatorial Pacific was established in 1979 during the Global Weather Experiment. From these buoys, simultaneous measurements were recorded of upper-ocean temperatures and currents along the equator at 165°E, 152°W and 110°W (Halpern 1980). Analyses of these observations demonstrated the need for long-term measurements of upper-ocean temperature and current fields as well as the feasibility of making reliable surface meteorological and upper-ocean observations for long periods. Theoretical ideas (McCreary 1976) about wind-generated downwelling equatorial Kelvin waves propagating 10000 km from the western Pacific to South America were confirmed by measurements from these moorings (Knox and Halpern 1982, Eriksen *et al.* 1983) and led to conceptual models for detecting the onset and termination phases of El Niño.

In the early 1980s, NOAA's EPOCS program initiated long-term surface-meteorological and upper-ocean current and temperature measurements on the equator using moored platforms. These measurements were designed to determine the annual cycles of the depth of the thermocline and the strength of the Equatorial Undercurrent (Halpern 1987c, Halpern and Weisberg 1989), to monitor the occurrence of Kelvin waves, and to validate the first generation of ocean general-circulation models for the tropical Pacific based on the primitive equations (Philander and Seigel 1985). Deepening of the thermocline, disappearance of the Equatorial Undercurrent, and appearance of the warm El Niño current running eastward along the equator—all features of the 1982–83 El Niño—were observed along the equator for the first time during an El Niño (Firing *et al.* 1983, Halpern *et al.* 1983, Halpern 1987c). As a result of these measurements, moored buoys were recognized as an essential feature of a long-term monitoring system for prediction and description of El Niño.

The undetected onset of the 1982–83 El Niño episode motivated real-time monitoring (Halpern 1996) of equatorial moored-buoy measurements. At the beginning of TOGA, moorings for making surface-meteorological and upper-ocean observations were established along the equator at 140°W, 125°W, and 110°W (Halpern 1988, Halpern *et al.* 1988). A fourth mooring was deployed at 165°E shortly thereafter. However, wind measurements at three or four sites in the equatorial Pacific were not sufficient to describe the surface wind field in the region extending from approximately 5°S to 5°N, where knowledge of surface winds is critically important for forecasting variations of the equatorial upper-ocean thermal and flow fields (McCreary 1976). Because the launch of a satellite-borne instrument for estimating vectors of equatorial Pacific surface winds was delayed for seven years, making it unavailable for TOGA, and because the idea of drifting thermistor chains did not seem feasible, an array of moored wind and subsurface measurements (Hayes et al. 1991), the TAO array, was contemplated.

The TAO array was not designed casually. Its design grew out of the need to observe the development of phases of ENSO in real time and to validate the forcing and responses of equatorial ocean models. The array was based on the developing knowledge of the scales of variability of the wind field and the effects of the wind field on the thermal structure of the upper ocean.

The first analysis of the sampling and accuracy requirements for tropical wind measurements was performed during the SEQUAL/FOCAL (Seasonal Equatorial Atlantic Experiment / Français Océan Climat Atlantique Equatorial) program in the Atlantic and published in the program plan for the SEQUAL Wind Program (SEQUAL 1982). Using the theoretical results of Cane and Sarachik (1978) for forced linear motion and assuming a random noise in the wind forcing, it was shown that monthly-mean wind stress needed to be known to about 0.1 Nm^{-2} over 500 km regions for dynamical models to reproduce thermocline displacements of 5 m. This simple result illustrated the crucial importance of accurate wind-stress information in the equatorial wave guide for quantitative tropical oceanography. It also prompted examination of the available data sets of surface winds and the resulting estimates of surface wind stress.

Soon after the SEQUAL/FOCAL efforts, plans were made for the Pacific Equatorial Dynamics (PEQUOD) experiment to study a number of different tropical phenomena. Wyrtki and Meyers (1975) had earlier published a climatology of surface winds that revealed the very large data voids typical of the historical data set for the Pacific region. There was strong variability between bimonthly periods and considerable noise in the individual vectors, which suggested that the atmosphere was strongly variable on these (or shorter) scales and that the sampling was not sufficient to resolve this variability. It became clear that obtaining an accurate wind field was necessary if models were to be used either for modeling tropical Pacific variability or as the basis for data-assimilation methods to help interpret observations. However, better information on the variability of surface winds in the tropics was needed to determine the appropriate scales and sampling strategies.

A study of 30 years of data from tropical Pacific islands collected by the New Zealand Meteorological Service was undertaken to determine the relevant spatial and temporal scales of variations in the surface winds. Luther and Harrison (1984) presented preliminary results comparing power spectra in the frequency domain. The spectra were based on the complete time series of monthly means as well as on time series of monthly means constructed from random sub-samples of the full time series. Luther and Harrison then compared their island-based results with the ship-based spectral results of Goldenberg and O'Brien (1981). The comparison suggested that away from the annual frequency, the ship observations were heavily aliased by unresolved high-frequency variability, and that about 30 observations per month were needed to resolve the dominant energetic scales of tropical wind variability. Analysis of the spatial coherence suggested that the coherence scales for the energetic frequency bands were about 2–3 degrees in latitude and 15 degrees in longitude. Harrison (1989) examined the effects of having limited zonal and meridional wind data in simulations of sea surface temperature during ENSO. This model study indicated that, within the model assumptions, it was necessary to know the wind stress all along the equatorial wave guide and within about 7 degrees of latitude of the equator for simulations of sea surface temperature to be accurate within about 0.5°C.

Stanley Hayes proposed the current configuration of the TOGA TAO array (Hayes *et al.* 1991). He based the design on the above-mentioned observational results of the coherence properties of the island winds and on the modeling results of the accuracy of the winds needed to accurately simulate the sea surface temperature. The planned array would span the equatorial wave guide, from 8°S to 8°N, and extend from the eastern Pacific all the way to the western Pacific. Although the instruments also measured surface temperature and subsurface thermal structure, the spacing and sampling design was based on the wind variability because so little was known about the spatial scales of temperature variability.

The typical TAO mooring, illustrated in Figure 6, is an ATLAS (Autonomous Temperature Line Acquisition System) mooring. It provides (via Service ARGOS) data on surface wind, sea surface temperature, surface air temperature, humidity, and subsurface temperatures at 10 levels down to 500 m. The toroidal buoy at the surface holds the atmospheric instruments and the satellite transmitter, while the line anchored to the bottom holds the thermistors and pressure devices that transmit their readings up the line. To save battery power, a certain amount of on-board processing is done, so that only hourly averages are transmitted. A tape keeps data with full temporal resolution. This tape is recovered when the buoy is removed and its replacement deployed, typically every six months. Also part of the TAO array are five equatorial stations that provide data on ocean currents. These moorings use a downward-looking acoustic Doppler current profiler (ADCP) to obtain measurements from the surface down to 250 m depth.

Implementation of the TAO array in the tropical Pacific did not begin until 1986, because the idea of a large TAO array was conceived only when it became clear that a satellite-based scatterometer for measuring surface winds

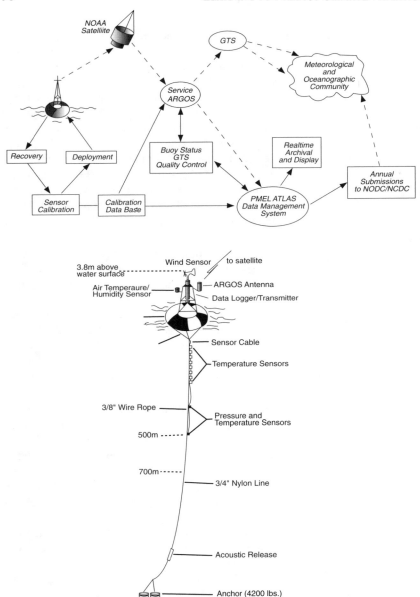

Figure 6. An ATLAS (Autonomous Temperature Line Acquisition System) Mooring. The lower diagram is a schematic of an ATLAS wind and thermistor-chain mooring. The upper panel provides an overview of the ATLAS data-transmission and archiving system. (Courtesy of M. McPhaden, NOAA/PMEL.)

would not be available for most of the duration of the TOGA Program. The array grew slowly. Gradual engineering improvements eventually allowed a predictable mooring lifetime of about six months. However, limitations of budgets and ship time, and the need for international cooperation, all delayed the completion of the planned array of moorings until very near the formal end of the program. The full complement of about 70 moorings was not completed until the end of 1994, as the TOGA Program ended. The growth of the array was aided by various process studies. Initially, moorings were deployed for EPOCS. Towards the end of the TOGA decade, moorings were deployed as part of the international COARE experiment, which required moorings in the western Pacific. Some western Pacific nations assisted in the deployment of these moorings.

The history of the TAO array provides a lesson in the interaction of process studies with monitoring networks. Many of the quantities now measured by the TAO array were originally expected to be measured by satellite. However, the TAO array was a necessary and creative response when the proposed satellite systems failed to materialize on schedule. Not only is the TAO system cheaper and better adapted to TOGA problems, but it also provides data on subsurface thermal structure, which could not be obtained directly (though some information can be inferred) from satellites.

The foundation for TOGA moored measurements was provided by the research community with research funding. Only now, with the end of TOGA, is this research observing system beginning its transition to a monitoring system in support of short-term climate predictions. Maintenance and operation of the array involve significant costs for ship time and communications, in addition to the costs of the mooring hardware. Mooring failures, data transmission problems and dropouts, mooring vandalism (especially in the western Pacific where the moorings are set in heavily fished areas), and the availability of ship resources to deploy and maintain the moorings are all continuing problems. Although moorings are inexpensive in comparison to spacecraft and satellites, a typical oceanographic vessel suitable for maintaining the TAO array costs $4–5 million per year. Securing reliable and sufficient international commitments of ship time—perhaps 1–2 ship-years per year—to sustain the array will require continued effort and attention in the years ahead. The TAO array is generally agreed to be the observational element that is most critical to sustain into the post-TOGA era, for two key reasons. First, it measures the high-priority fields (sea surface temperature, surface winds, and upper-ocean thermal structure) in otherwise inaccessible locations, relatively economically. Second, the TAO array has reached full deployment only recently, so there has not yet been time to make a careful assessment of its impact on predictions, particularly predictions of ENSO phenomena (NRC 1994a). Extensions of this type of observing system into higher latitudes, or into the Indian and/or Atlantic Oceans, while clearly desirable for predicting short-term climate variations, will require detailed study, justification, and funding.

Satellite Observations

No satellite sensor or spacecraft platform can be considered to be an accomplishment of the U.S. TOGA Program, although TOGA did drive improvements in some systems and their use. This is in contrast to the *in situ* observing platforms that form the basis for the TOGA Observing System. Notwithstanding this difference, remotely sensed observations of the coupled ocean–atmosphere system played a key role in the development and evolution of the TOGA Program. For example, composite pictures of cloudiness from the satellites ESSA-3 and ESSA-5 were used by Bjerknes (1969), to formulate the hypothesis that ENSO resulted from a coupling of oceanic and atmospheric circulations.

In the early 1980s, a NASA working group was charged with identifying areas of ocean science that would experience significant advancement as a result of remotely sensed measurements of the surface wind field. That group singled out studies of the El Niño phenomenon as the major beneficiary (O'Brien *et al.* 1982). At the outset of TOGA, the proposed 1990 launch of a NASA scatterometer (NSCAT) on board the U.S. Navy Remote Ocean Sensing System (NROSS) satellite was expected to provide data on winds at the sea surface. It was anticipated that these data would have a dramatic impact on modeling and assimilation studies during the second half of TOGA. To the dismay of many in the TOGA community, the NROSS mission was canceled, and NSCAT was not launched until August 1996 on board the Japanese Advanced Earth Observing Satellite (ADEOS). The lack of comprehensive satellite coverage of the equatorial Pacific surface-wind field played a major part in the justification and deployment of the TAO array.

Limited observations of the surface-wind field became available in 1978 from Seasat and from Nimbus SMMR (Scanning Multichannel Microwave Radiometer). Wind-speed estimates from the passive microwave sensor on board the Defense Meteorological Satellite Program (DMSP) SSM/I (Special Sensor Microwave/Imager) satellite became routine in 1987 (Wentz 1989). Global coverage with 35 km resolution is now provided approximately every 1.5 days by the two satellites in orbit as part of this operational series. Various techniques have been proposed to convert this wind-speed information into wind velocity (e.g., Atlas *et al.* 1991, Wentz 1992). Examination of these wind fields for the tropical Pacific Ocean indicates that they are comparable to existing operational and subjectively derived wind fields (Busalacchi *et al.* 1993) and direct wind-speed measurements from the TAO array. Remotely sensed observations of the surface-wind velocity field began in July 1991, when the

European ERS-1 (Earth Resources Satellite) was launched with a radar scatterometer on board. The processing of these data has since undergone a number of refinements (Stoffelen and Anderson 1992, Freilich and Dunbar 1993).

The launch of the U.S. Navy's Geosat in 1985 made possible the use of satellite altimetric measurements to obtain near-global sea-level coverage. When the Geosat data were declassified in 1986, they increased, rather than reduced, the value of the *in situ* sea-level observations. Combined, the two data sources provided a unique view of equatorial wave processes that had been sparsely observed by the *in situ* network. With the launch of TOPEX/Poseidon in 1992, *in situ* data revealed the high accuracy of the space-based measurements, and gave confidence to their interpretation. The interpretation of the long time series of *in situ* data can also be enhanced using the spatial coverage provided by the satellite.

During the course of TOGA, remote measurements of a number of quantities—e.g., sea surface temperature, sea level, water vapor, and cloud fraction—were obtained from both research and operational satellites. Unfortunately, the TOGA Program ended prior to rigorous assessment of the ultimate value to ENSO monitoring and prediction of many of the remotely sensed quantities. It is left to the emerging GOALS Program to accomplish this assessment.

Observing the Atmosphere and Ocean for TOGA

TOGA scientists recognized early, on the basis first of the work of Bjerknes (1969) and later of the work of Zebiak and Cane (1987), that ENSO was a coupled atmosphere–ocean phenomenon. The prime quantities of observational interest were sea surface temperature, surface wind velocity, and upper-ocean thermal structure. The identification of these prime quantities enabled TOGA to set its priorities clearly and unambiguously. Other quantities, such as sea surface salinity, surface heat fluxes, and sea-level height, were all of interest but of a lower priority.

Sea Surface Temperature

Sea surface temperature is the prime oceanic quantity to which the atmosphere in the tropics responds. Sea surface temperature is measured *in situ* by the TAO array, the VOS network, and the Global Drifter Array. A great challenge has been to use these relatively sparse (except perhaps close to the equator) measurements to make gridded global fields of sea surface temperature. A combination of remote and *in situ* data has proved to be the solution.

The definition of sea surface temperature also presents problems. Ships and buoys vary in the exact depth, water volume, and averaging time used when measuring "bulk" *in situ* sea surface temperatures. These variations in technique have made it very difficult to create long-term homogenous time series. Satellite measurements integrate over a large area but see only the "skin temperature" from a layer less than 10^{-4} m in thickness. Skin temperature can often be a few tenths of a kelvin less than bulk temperature. Under low-wind conditions during TOGA COARE, skin temperature was observed to be as much as 4 K higher than the bulk temperature.

The large-scale monitoring of sea surface temperature using infrared satellite measurements from the operational AVHRR provided a critical contribution throughout the entire TOGA Program. These measurements were the principal source of information going into the sea surface temperature indices that were used to validate coupled atmosphere–ocean prediction models. They served as the basis for the fields of sea surface temperature used to force simulations with atmospheric general-circulation models, and to evaluate simulations with ocean general-circulation models. However, transformation of satellite measurements of radiances into sea surface temperature is not straightforward, because of calibration biases and contamination from clouds and atmospheric aerosols, such as those from volcanic eruptions.

In view of the limitations of both satellite and *in situ* data, TOGA investigators made extensive use of blended analyses of sea surface temperature produced by the U.S. Climate Analysis Center (CAC, now the Climate Prediction Center, CPC). In these analyses, the biases of the AVHRR data were reduced in an objective manner by including measurements of sea surface temperature from the Global Drifter Array and VOS surface observations. Towards the end of TOGA, a new optimal interpolation scheme that improved on this blended product was developed at the National Meteorological Center (NMC, now the National Centers for Environmental Prediction, NCEP) (Reynolds and Smith 1994). At the end of TOGA, the CAC product was available weekly on a global 1°×1° grid. The root-mean-square error between this product and independent measurements of sea surface temperature from the TAO array was 0.3–0.7 K. The confidence placed in this product attests to the importance of combining the spatial coverage provided by satellite measurements with the high accuracy of *in situ* measurements.

Surface Wind and Stress

Wind stress at the surface of the equatorial ocean is the prime driver of oceanic dynamics and a crucial determinant of sea surface temperature. It therefore shared with sea surface temperature the highest observational priority in TOGA. The need for accurate wind-stress estimates has been known since equatorial ocean models became good enough for detailed comparisons with observations. The long time series of hand-analyzed tropical Pacific pseudo-stresses (1963 to date) produced at Florida State University has proven invaluable in running ocean models, although inadequacies in these estimates of the wind fields have long been recognized (see, e.g., Halpern 1988, Legler 1991).

The main source of equatorial surface-wind observations was the TAO array and the VOS network, because of the sparseness of islands in the Pacific. Both of these sources of observations reported immediately so that the data were available to the operational weather prediction centers (especially the U.S. NMC and the European Centre for Medium Range Weather Forecasting (ECMWF)). Outside the equatorial region, the main sources of wind data are the VOS network and satellite-based scatterometers.

Comparison of scatterometer data with monthly-mean wind speeds from TOGA TAO array for 1992 yielded cross-correlations of approximately 0.8 and root-mean-square differences of 1.4 m/s. The implications of this result for long-term monitoring, and for evaluating the advantages and possible redundancies of the suite of wind information sources—such as passive microwave sensors, radar scatterometers, and the *in situ* measurements from the TOGA TAO array—have not yet been determined.

Subsurface Temperature

Subsurface thermal structure determines the location of the thermocline and therefore the location of the water destined to interact with the surface. Knowledge of this thermal structure is crucial to all descriptions and predictions of sea surface temperature. The initialization of models for making ENSO predictions requires specification of the subsurface thermal structure because the ocean's evolution on seasonal-to-interannual time scales is controlled, in part, by the planetary waves evident in the thermal structure.

The VOS network and the TAO array supply the only regular observations of subsurface temperature. Only from the TAO array are gridded fields of subsurface temperature available immediately. Also, the VOS data are irregular in space and time, and contaminated by higher-frequency waves that cannot be eliminated with the use of a single measurement at each location. However, in regions without moorings, the VOS system is invaluable for providing subsurface temperatures.

Elements of the TAO array have been in place sufficiently long for the gradual development of climatologies of subsurface temperature. Figure 7 shows the climatology of currents and subsurface temperatures on the equator at 110°W. Long-time-scale phenomena contribute to the mean annual cycle. Therefore, to derive climatologies for comparison with the annual cycles simulated by coupled atmosphere–ocean models, long deployments of observing systems are needed. Measurements must also be dense in time because shorter time-scale phenomena can alias the results.

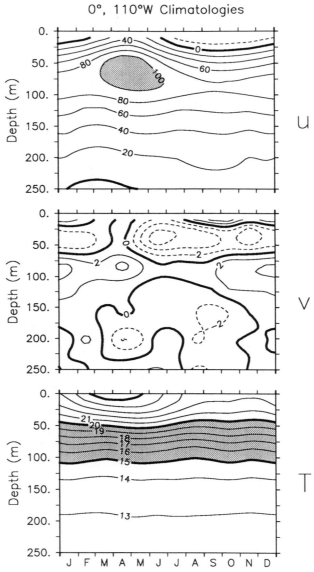

Figure 7. Climatology of the near-surface equatorial ocean at 110°W. Shown as a function of depth and month are (top panel) meridional velocity, u, in cm/s; (middle panel) zonal velocity, v, in cm/s; and (bottom panel) temperature in degrees Celsius. For u, dashed contours indicate westward flow, and for v, dashed contours indicate southward flow. (From McPhaden and McCarty 1992.)

Sea Level

Sea level can be an important tool in the study of heat budgets over large regions in the tropics because sea level is affected by changes in subsurface heat content. Wyrtki (1985) created an index of warm water in the equatorial Pacific for use in monitoring the development of El Niño. The global network of operational tide gauges provided valuable information on large-scale sea-level fluctuations for TOGA and other purposes. In the TOGA context, sea-level fields provide a tool for model validation and serve as a useful integral constraint for model initialization.

Sea-level differences can be interpreted in terms of the changing strength of ocean currents because the topography of the sea surface is linked by geostrophy with ocean circulation. This linkage was exploited by Wyrkti (1974, 1987) to examine equatorial currents in the Pacific. Sea level is a measure of other oceanographic quantities of interest as well, such as heat storage or thermocline depth. It is especially useful for estimating thermocline depth in the tropics, where the density structure can be approximated by a two-layer system (Rebert *et al.* 1985).

Sea level rises when the thermocline deepens and falls when the thermocline shoals. Therefore, it serves as a marker of the oceanic thermal structure. In the tropics, the interannual range of sea level is a few tens of centimeters.

Tide gauges that are part of the TOGA Sea Level Network provide most of the measurements. It is difficult to obtain a picture of the evolving thermal structure of the Pacific from the Sea Level Network alone, because this network is limited to coastal stations and islands, and the islands are poorly distributed throughout the interior of the Pacific. Sea level estimates may be obtained by integrating the thermal structure of the ocean; it can thus serve as an integral constraint on numerical models of the ocean.

Measurements of sea surface topography from satellite altimeters were available from three different missions during TOGA—Geosat (1985–1989), ERS-1 (July 1991 onward), and TOPEX/Poseidon (August 1992 onward). As time progressed, the measurement accuracy of these satellite altimeters improved significantly. Globally averaged root-mean-square errors in determinations of both orbit radius and the corrected values of surface height dropped from approximately 30 cm and 8 cm respectively for Geosat, down to 20 cm and 6 cm for ERS-1, down to 4 cm and 3 cm for TOPEX/Poseidon. The errors are now sufficiently small that when the data are used to study the interannual variations of sea level in the tropical ocean (of order tens of centimeters) or the annual variability (of order centimeters), very little, if any, post-processing of the altimeter data is required.

Although the Geosat mission was originally intended for geodetic and mesoscale-oceanography applications of the U.S. Navy, the altimetric data proved surprisingly useful for monitoring equatorial-wave propagation, once large-scale orbit errors had been removed (Miller et al. 1988, Delcroix et al. 1991). The Geosat data have been used to track the space-time progression of a number of eastward-propagating Kelvin-wave pulses across the equatorial Pacific basin in response to westerly winds west of the dateline during the 1986–1987 ENSO warm phase (Miller et al. 1988, Cheney and Miller 1988). Beyond the tracking of propagating wave features, the meridional curvature (second spatial derivative) of the sea level across the equator has been used to infer changes in the zonal current field (Picaut et al. 1990). The resulting altimeter-derived estimates of the geostrophic flow field have been shown to agree well with the low-frequency, near-surface zonal current observed at the three TOGA current-meter moorings in the western, central, and eastern equatorial Pacific.

Prior to the launch of TOPEX/Poseidon, it was anticipated that the increased accuracy of the instruments on that three- to five-year mission, relative to that of previous altimeters, would be able to capture the sea-level evolution of a complete ENSO cycle. As it turned out, the timing of the launch in August 1992 coincided with middle of the protracted 1991–1992–1993 warm event. Comparisons of the first year of altimeter data from TOPEX with estimates of dynamic topography from more than 60 moorings of the TAO array indicate that the low-latitude sea-level variability in the Pacific for this period can be attributed primarily to equatorial Kelvin wave activity. These Kelvin waves appear to be generated west of the dateline by intense wind bursts that occur in association with the warm event (Busalacchi et al. 1994). Cross-correlations between data from the satellite altimeter and data from the moorings were generally greater than 0.7, with root-mean-square differences of approximately 4 cm. Comparisons of the satellite-altimeter data with sea-level data from approximately 70 island tide-gauge stations (primarily in the tropics) yielded similar results, with an average root-mean-square difference of 4.3 cm and a cross correlation of 0.66 for time scales greater than 10 days (Mitchum 1994).

A highlight of the TOPEX/Poseidon mission has been the use of altimetric data to track the sea-level signal of ENSO during 1992–1993 (CEES 1993). This success comes at a time when, as is the case with satellite scatterometry, the ultimate value of satellite altimetry for ENSO monitoring and prediction has yet to be determined. The next launch of a TOPEX-class altimeter is scheduled, with some uncertainty, to be NASA's Earth Observing System Altimeter mission in about 2000. A one- to two-year gap in sea-level monitoring with this class of altimeter is anticipated. At some future time, regular altimetric data from satellites may relax the need, at least for studying ENSO, for a spatially dense tide-gauge network, such as the one developed in the Pacific.

Heat and Moisture Fluxes

Heat and moisture fluxes between the ocean surface and the atmosphere are a major influence on sea surface temperature. Throughout most of the tropics, the sensible heat flux from the ocean into the atmosphere is relatively small (on the order of 10 Wm^{-2}) and relatively unchanging. The basic heat balance at the tropical sea surface is between the net radiation and the evaporation, with the heat flux into the ocean determined as the difference. The net heat flux into the ocean can be as large as 100 Wm^{-2} in the cold tongue of the eastern Pacific or as small as 10 Wm^{-2} in the Pacific warm pool. Evaporation is especially difficult to measure. It is often parameterized as a product of the wind speed and the difference in specific humidity between the surface air and 10 m up. The heat flux associated with latent heat of evaporation is generally known within only about 30 Wm^{-2} because of errors in measurement of both the winds and the specific humidities. Improved bulk-flux algorithms, especially for light-wind conditions, were developed by using direct measurements of evaporation taken during COARE (Fairall *et al.* 1996; Bradley and Weller 1995a, b).

Radiances measured from satellites have proven useful in deriving estimates of latent heat flux, radiative fluxes, and precipitation. Although data from satellites have been used to produce these flux fields, only limited validation of these fields in the tropics using *in situ* data has been performed. The measurements made during COARE are proving invaluable for calibrating satellite-based measurements.

Passive-microwave data can be used to estimate water-vapor distributions because the microwave spectrum is sensitive to the presence of water molecules. Using passive-microwave data, the relation of water vapor to sea surface temperature and convection during ENSO has been studied (Prabhakara *et al.* 1985, Liu 1986). Total-column water vapor, which can also be estimated from passive microwave data, has been used by Liu (1986) to derive an empirical formula for monthly-mean ocean-surface humidity. Combining this result with the passive-microwave estimates of the surface wind speed, Liu (1988) was able to estimate changes in the latent-heat flux over the tropical Pacific Ocean during the 1982–1983 warm phase of ENSO.

Additional approaches to deriving the fields of surface fluxes utilize atmospheric general-circulation models. One approach derives the fluxes from the operational global analyses performed at the various global weather-forecast centers. Another uses the output of an atmospheric general-circulation model forced by observed sea surface temperatures. In this latter approach, to the extent that the model is valid, the surface fluxes into the atmosphere (and therefore the residual heat flux into the ocean) are known. While this method is difficult to validate (the COARE data set should be extremely useful in this regard), it offers a path to gradually improving our knowledge of the fluxes (see, for example, Weller and Anderson 1996). It can also be applied retrospectively as long as the sea surface temperature is sufficiently well known. Reanalysis of atmospheric data sets using atmospheric general-circulation models offers the possibility of preparing long time series of surface fluxes.

Upper-Air Observations

The distribution and availability of upper-air observations were a major concern during TOGA, and will continue to be a major concern for climate research. The overwhelming majority of upper-air observations derive from the Basic Synoptic Network (BSN) of the World Weather Watch (WWW). Some 400 BSN stations lie in the tropical belt (30°S–30°N), which was of primary interest to TOGA. TOGA identified 182 of these stations as having the highest priority because of the need for global coverage in the tropics and for data in areas where few other measurement platforms exist.

Unfortunately, few of these highest-priority stations report regularly and promptly (i.e., in real time via the Global Telecommunications System). Only those data that are transmitted promptly can be used in operational analyses and other products. Furthermore, the BSN has been deteriorating. The inadequacies of the WWW caused serious problems for TOGA. Successor efforts to understand and predict seasonal-to-interannual variations of the physical climate system will face similar problems. Solutions will have to involve a much broader constituency than just the climate research and prediction community, because the WWW is founded on concerns for operational weather forecasting and derives support from national budgets for this activity. Still, the climate community can lend its voice in support of this essential observing system.

Upper-air observations are also obtained using Doppler wind profilers (Gage *et al.* 1994), aircraft en-route reports (AIREPs), and estimates of winds made from cloud drifts shown in satellite images. Doppler profilers exist or are under construction at seven sites in the tropical Pacific, with two additional sites under consideration; all were developed as specific contributions for the TOGA program. Wind profilers can form a partial potential solution to the difficulties of maintaining manned BSN/WWW stations because they can operate unattended for long periods of time. Automation of the transmission of AIREPs, via Meteorological Data Collection and Reporting Systems (MDCRS) or similar techniques, could greatly enhance the receipt and usefulness of wind data from civil aircraft, which are usually of high quality. Winds estimated from cloud drift are an established and useful data set for studies of certain regions, including much of the tropics. These wind estimates can be made only as long as the geostationary satellites and the associated image-analysis effort are sustained.

Other Quantities

Measurements from the AVHRR, a satellite-based infrared imager, were used to study a number of attributes of ENSO in addition to sea surface temperature. Monthly anomalies of the outgoing long-wave radiation (OLR) observed by the AVHRR during TOGA proved to be an effective qualitative measure of tropical convection, and these measurements also served as a proxy for precipitation. These OLR estimates were used to monitor the extreme shifts in regions of strong convection during ENSO and the resulting surpluses or deficits, compared to normal conditions, in precipitation. More quantitative estimates of monthly precipitation totals are expected from the Global Precipitation Climatology Project (GPCP). Begun in 1987, this project is using infrared and passive microwave measurements to provide monthly mean estimates of precipitation with $2.5°\times2.5°$ resolution over the entire globe for 1986–1995 (WMO 1990). A great leap in the quality of satellite-based precipitation estimation is expected with the launch of the Tropical Rainfall Measuring Mission in 1997. A suite of visible, infrared, passive-microwave, and active-microwave radiometers will be dedicated to measuring precipitation, and also the vertical profile of the latent heat released in the tropical atmosphere.

The AVHRR instrument and visible and infrared channels on a number of geostationary satellites are being used with sounding data from satellites to map cloud fraction and cloud height. As part of the International Cloud Climatology Project (Rossow and Schiffer 1991), these data are being used to infer atmospheric and surface radiation properties. Data are now available as far back as July 1983, and are likely to stimulate retrospective analyses of the role of clouds and radiation during the TOGA decade. Moreover, the availability of net shortwave radiation fields, together with observations of ocean color expected from the SeaWifs mission, will permit an evaluation of how penetrating radiation may affect upper-ocean temperature variability (Lewis *et al.* 1990).

Several remote-sensing techniques hold promise for the long-term large-scale monitoring of key quantities of the climate system (see Gurney *et al.* 1993, especially, within that volume, K.-M. Lau and Busalacchi 1993). Over the next decade, a number of national and international earth-observing satellite missions are expected to provide unprecedented concurrent estimates of wind velocity at the ocean surface, wind speed at the ocean surface, columnar water vapor, sea surface height, tropical rain rates, sea surface temperature, albedo, cloud fraction, surface irradiance, and ocean color. However, experience has shown that the promise and potential of timely remotely sensed observations can go unfulfilled for reasons of cost, schedule, complexity, or lack of strong advocacy.

PROCESS STUDIES

> Process studies are focused efforts of limited duration, and usually field experiments with limited geography, designed to collect data for the purpose of understanding particular physical processes. Most of the process studies associated with TOGA concentrated on the dynamics of either the atmosphere or the ocean. The massive COARE study was the one study that truly looked at the coupled system of the atmosphere and ocean. COARE was designed to develop understanding of the evolution of the warm pool of the tropical Pacific and its concomitant precipitation. The experiment was also designed to develop improved parameterizations of the physical processes that couple the atmosphere and ocean in the tropical Pacific.

At the outset of the TOGA Program, several conceptual models of the ocean and atmosphere were used to explain the interannual warming of the eastern tropical Pacific. A number of limited-duration experiments were conducted to determine the relative importance of several processes and to improve the representation of those processes in models. These experiments aimed to reduce the uncertainties associated with numerical simulations and predictions of month-to-month fluctuations of sea surface temperature, especially in the eastern tropical Pacific. Process studies have improved our understanding of the physical mechanisms associated with ocean–atmosphere interactions, yielded improved parameterizations of processes having time and space scales smaller than those associated with models, provided critical tests of these models, and contributed to the evolution of the TOGA Observing System.

Nearly all process studies for TOGA were performed in the Pacific, because of the strong focus on ENSO. Some process studies were supported almost entirely by the U.S. TOGA Program, while others were supported by a broad international effort. Most of the process studies conducted during the first half of TOGA addressed the physics of either the ocean or the atmosphere. Only COARE, begun in 1992, addressed the physics of the coupled system. Process studies intrinsic to the coupling of the oceans and atmosphere are discussed here first, followed by atmospheric studies, and then oceanic studies.

Coupled Ocean–Atmosphere Response Experiment (COARE)

The TOGA Coupled Ocean–Atmosphere Response Experiment (COARE) was conducted to study the strong air–sea interactions in the western equatorial Pacific Ocean, where sea surface temperatures are warmer than 29°C and where deep convection and heavy precipitation occur (Webster and Lukas 1992). This region is probably the most important one for the development of ENSO cycles. The Intensive Observation Period (IOP), from 1 November 1992 to 28 February 1993, was embedded in a period of enhanced monitoring from September 1991 to May 1994 (see Figure 8). TOGA COARE was an internationally endorsed addition to the International TOGA Program (WCRP 1990), and the substantial

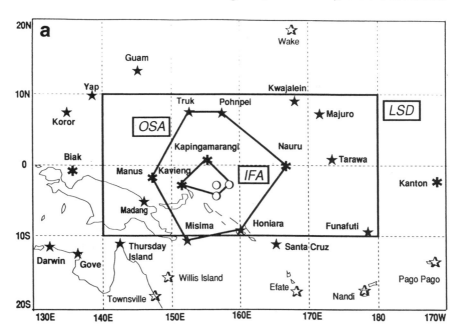

Figure 8. Composite structure of the Intensive Operation Period for TOGA COARE. The legend shows the symbols used to represent the different type of observational platforms. Various domains are labeled: the entire COARE domain (a), the Large-Scale Domain (LSD), the Outer Sounding Array (OSA), and the Intensive Flux Array (IFA). (From Webster and Lukas 1992, reprinted by permission of the American Meteorological Society.)

resources required for its implementation were contributed by nineteen nations, including the United States. Details on the contributions from Australia, China, the Federated States of Micronesia, France, Germany, Indonesia, Japan, Nauru, New Zealand, Papua New Guinea, the Philippines, the Republic of Korea, Russia, the Solomon Islands, Taiwan, the United Kingdom, and the United States can be found in WCRP 1993.

The objectives of COARE were to answer several questions:

- What is the relationship of synoptic and mesoscale air–sea interaction in the warm-pool region to the large-scale, low-frequency behavior of the coupled ocean-atmosphere system?
- How are air–sea fluxes of heat, moisture, and momentum in the warm-pool region modulated by synoptic and mesoscale atmospheric and oceanic forcing?
- What are the structures and morphology of the synoptic and mesoscale components over the warm pool? and
- What is the upper-ocean response of the warm pool to heat, moisture, and momentum fluxes associated with synoptic and mesoscale atmospheric systems?

TOGA COARE was motivated by:

- the difficulty that coupled ocean–atmosphere models have in adequately simulating the mean state and variability of the thermal and flow fields in the lower atmosphere and upper ocean, because of large uncertainties in the components of the net air–sea flux, onset of atmospheric convection, and ocean mixing parameterization;
- a dearth of high-quality measurements of air–sea fluxes of heat, moisture, and momentum in the western tropical Pacific;
- uncertainty about the importance of the influence of rainfall upon buoyancy and mixing in the upper ocean;
- concerns about the interpretation of space-based observations of sea surface temperature and surface wind; and
- multiple-scale interactions that extend the oceanic and atmospheric influence of the western Pacific warm-pool area to other regions and vice versa.

Because of the size and complexity of TOGA COARE, a TOGA COARE International Project Office (TCIPO) was established in Boulder, Colorado, distinct from the Geneva-based International TOGA Project Office (ITPO). The TCIPO was charged with the daunting task of providing the logistical support required for the ground, air, and sea deployment of a multinational field campaign in a relatively remote portion of the world. U.S. resources for COARE were provided by NOAA, NSF, NASA, the Office of Naval Research (ONR), and the Department of Energy (DOE). Of significant note were the successful efforts by the agencies to coordinate different agency missions and requirements to the benefit of COARE. An International TOGA COARE Data Workshop was held in Toulouse during August 1994 to address the multidisciplinary issues intrinsic to the objectives of the project.

The basic strategy for TOGA COARE was to obtain comprehensive atmospheric and oceanic data sets to guide the improvement and evaluation of coupled models. The 4-month IOP was embedded within a 2.5-year enhancement of the relatively sparse TOGA Observing System in the western equatorial Pacific. Observations were combined with model-based data assimilation to achieve the best description of the dynamics of the interactions between the ocean and atmosphere. Observations were nested in space to relate processes occurring on different scales.

The IOP was set for the northern-hemisphere winter (November 1992 to February 1993) to maximize the probability of strong westerly wind events, which were expected to be the periods of most intense air–sea interaction. Furthermore, November to February is a time of active transition during ENSO, while also being a period of the maximum strength of the east Asia winter monsoon. Leading up to the IOP, conditions in the tropical Pacific had relaxed from warm to near normal, with slightly cold surface waters in the eastern equatorial Pacific (see Figure 9). However, a cold event did not materialize and there was a return to warmer-than-normal surface waters in the central equatorial Pacific. TOGA COARE was conducted during the redevelopment of the warm phase of ENSO that had begun in 1991, allowing an unprecedented sampling of the conditions during a re-intensification of El Niño (Gutzler *et al.* 1994, Lukas *et al.* 1995).

COARE's strong emphasis on the measurement of air–sea fluxes was embodied in the Intensive Flux Array (IFA). Measurements of fluxes using eddy-correlation techniques were made from ships and from aircraft. Ship-, satellite-, and aircraft-based measurements also provided indirect estimates of fluxes. Doppler weather-radar measurements yielded high-quality estimates of rainfall over broad areas with high resolution. Dual-Doppler-radar measurements (measurements using two radars, making it possible to determine the horizontal wind speed and direction) from ship and aircraft provided detailed momentum-flux estimates for limited periods.

The large dynamic range of atmospheric and oceanic conditions observed during TOGA COARE confirmed hypotheses about the role of air–sea interaction in the warm pool and provided an excellent data set for model development. Strong intraseasonal atmospheric oscillations occurred, along with some episodes of westerly winds, including a strong westerly wind burst over the IFA

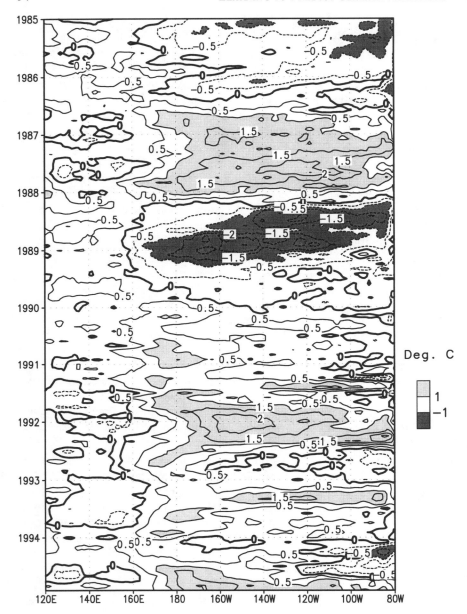

Figure 9. Monthly-mean anomalies of sea surface temperature averaged over 5°S to 5°N during the TOGA years. Anomalies are based on the adjusted optimal-interpolation climatology of Reynolds and Smith (1994). (Courtesy of M. Halpert, NOAA.)

between 20 December 1992 and 3 January 1993. The burst generated an eastward-moving equatorial Kelvin wave, which caused the thermocline to deepen and the sea surface to warm across the equatorial Pacific Ocean. It directly forced the warm surface waters to move eastward. Also, the burst cooled the western Pacific through air–sea heat fluxes and upper-ocean mixing. The net effect was to displace the warm pool eastward.

Observations of rainfall over the ocean were made using Doppler weather radars on ships and on research aircraft. These observations provided the elusive areal averages of precipitation needed to compute the net air–sea flux of fresh water, as well as the detailed motions inside atmospheric convective systems. A new finding was that roughly 15 percent of the heavy rainfall over the warm pool occurs during shallow convection, which recycles local moisture rather than coming from moisture convergence on larger spatial scales. This observation of so-called "warm rain" has important implications for modeling the dynamic response of the atmosphere on large scales to convection in this region, and may be related to the inversion near the 0°C level (Johnson et al. 1996, Lin and Johnson 1996a). The observed IOP-average precipitation of 6 mm/day over the IFA (see Figure 8) agreed well with the average computed from the moisture budget derived from atmospheric soundings. The observed variability of precipitation compared reasonably well among these direct measures and with estimates based on SSM/I. However, estimates of precipitation variations from the GOES (Geostationary Operational Environmental Satellite) Precipitation Index differed substantially from the observations (Lin and Johnson 1996b).

Special time periods for instrument intercomparison helped to provide confidence in the accuracy of the fluxes. When estimates of heat flux were compared with changes in the heat content of the upper ocean, it appeared that the local budget of surface heat flux might be closed to within 10 Wm^{-2} for time periods of days to weeks, although contributions from horizontal advection were clearly important at certain times. An algorithm for improved estimates of heat fluxes using a bulk formula was developed (Fairall et al. 1996). Mesoscale atmospheric circulations associated with convection and precipitation were seen to modulate the air–sea heat flux.

Heavy rainfall events were found to be responsible for salinity and temperature changes that substantially modify upper-ocean mixing, and thus the heat and momentum fluxes. During the strong westerly wind burst of December 1992, the mixed layer deepened and cooled. When the burst ceased, heavy rains caused the mixed layer to become very thin, allowing it to warm very rapidly to temperatures near to those prevailing prior to the burst. However, the reduction of the mixed-layer thickness caused by the rains accompanying the westerly wind burst remained for a substantial period after the burst, leading to enhanced sensitivity of sea surface temperature to subsequent wind events. Strong eastward surface flow forced by the westerly wind burst enhanced the vertical shear, which sustained mixing below the surface mixed layer for several days after the burst had ended (Smyth *et al.* 1996a, b), resulting in an even greater thermodynamic impact of the westerly wind burst on the warm-pool region.

Strong diurnal warming of the upper few meters of the ocean, sometimes as much as 4 K, was observed (Soloviev and Lukas 1996, Webster *et al.* 1996). This diurnal variation can lead to significant aliasing of sea surface temperature time series unless sampling takes place sufficiently frequently throughout the day. Also, the diurnal cycle increases the difference between skin and bulk sea temperatures. Caution is therefore necessary in the interpretation of climatological and satellite sea surface temperature data. Sampling frequency for satellite-derived sea surface temperature by infrared instruments is reduced by the abundant cloudiness of the warm-pool region, leading to potential biases.

Short-wave radiation penetrates to depths below the mixed layer, contributing significantly to heat content variations there, which could influence sea surface temperature and create temperature inversions. During the strong westerly wind burst, entrainment of nutrients into the mixed layer was associated with a phytoplankton bloom that, by increasing the turbidity, decreased the loss of penetrating solar radiation from the mixed layer, leading to an additional warming of about 0.1 K per month (Siegal *et al.* 1995).

While the results of the COARE experiment are still being analyzed and most of the work has not yet been published, some highlights have already emerged. The strategy of combining state-of-the-art turbulent-flux measurements on a few ships and aircraft, embedded in an observational array with high-quality *in situ* and remotely-sensed bulk measurements was successful. After a careful comparison of turbulence and bulk observations, made specifically to detect and resolve measurement problems, a revised bulk turbulent-flux algorithm has been developed (Fairall *et al.* 1996). This parameterization improves flux estimates under low-wind conditions, an important objective of COARE.

Prior to COARE, it was thought that the tropical atmosphere could sustain intraseasonal (40–60 day, or Madden-Julian) oscillations without any interaction with the ocean. Now, with COARE observations in hand, it is clear that the intraseasonal oscillation is a coupled phenomenon. The oceanic processes are highly correlated with the atmospheric processes, and phased so as to provide important feedbacks. These feedbacks have strong influences on the strength and propagation characteristics of the oscillations.

COARE observations and numerical simulations confirmed an hypothesis that the hydrological cycle is thermodynamically important for the upper ocean. Changes in local precipitation can alter the mixed-layer depth and thus alter the rates of change in sea surface temperature during warming and cooling periods in the warm pool (Smyth *et al.* 1996b, Anderson *et al.* 1996). Increased rain rates, or reduced wind forcing, yield a shallow mixed layer and may lead to a net cooling of the sea surface, in spite of a net-positive heat flux because of loss of penetrating short-wave radiation through the base of the mixed layer. Decreased rain rates, or increased wind forcing, can initiate entrainment cooling of the surface mixed layer. Furthermore, the stabilization of the surface layer by rainfall leads to a shallower layer on which a given wind stress acts, thereby leading to larger ocean-surface currents (Weller and Anderson 1996) than would exist in the absence of rainfall.

Research aircraft and seagoing vessels equipped with Doppler radar collected measurements allowing description of a broad spectrum over the warm pool of precipitating clouds, ranging from isolated showers dominated by warm rain processes to huge "superclusters" reaching the tropopause. These data describe the mesoscale precipitation structure and momentum-flux characteristics of the atmospheric convection in both space and time (Mapes and Houze 1993, Chen *et al.* 1996). In tandem with carefully calibrated *in situ* thermodynamic measurements, these data sets will allow greatly improved estimates of the effects of organized atmospheric convection on air–sea fluxes to be made (see, for example, Young *et al.* 1995).

Central Pacific Experiment (CEPEX)

The Central Pacific Experiment (CEPEX) was conducted for one month (March 1993) following the TOGA COARE IOP (CEPEX [no date]). Its timing was motivated by a desire to utilize COARE resources already in the western Pacific and also to extend the COARE observations. CEPEX was designed to investigate hypotheses on the regulation of sea surface temperature. Two of these hypotheses are that cooling from evaporation limits sea surface temperature (Newell 1979,1986) and that the reduction of incoming solar radiation by cirrus clouds is an important negative feedback to sea surface temperature (Ramanathan and Collins 1991). Improved understanding of various cloud feedbacks is particularly important for the development of the global climate models needed for the prediction of climate variability. CEPEX was a U.S. experiment funded by the NSF, DOE, NASA, and NOAA.

CEPEX was conducted in the central Pacific, to the east of the COARE IFA, between approximately 160°E and 155°W. It used a variety of platforms, including ships, aircraft, satellites, and the TOGA TAO array, to measure radiation and moisture, as well as the physical quantities that determine the response of sea surface temperature to radiative changes in the atmosphere.

Atmospheric soundings during CEPEX indicated that deep convection sharply increases mid- to upper-tropospheric moisture, with an enhancement of the greenhouse effect closely related to synoptic-scale convective events (Weaver et al. 1994). The enhanced greenhouse effect leads to significant radiative heating of both the surface and the atmospheric column by reducing the net outgoing flux of radiation. Upper-tropospheric mesoscale convective clouds contribute most of the satellite-derived cloud reflectivity in the central and western Pacific (Collins et al. 1996). Models for closure of the energy budget of the warm pool indicate that the effect of deep convection on surface insulation may be much larger than estimated by current general-circulation models (Ramanathan et al. 1995). Evaporation, computed from moored-buoy measurements, was anticorrelated with sea surface temperature in the western Pacific because of the light winds (Zhang and McPhaden 1995). The CEPEX data indicate that evaporative cooling is not sufficient to limit sea surface temperature and they do not rule out the importance of the cirrus feedback. A summary of the status of the debate on regulation of sea surface temperature has been provided by Waliser (1996).

Australian Monsoon Experiment (AMEX) / Equatorial Monsoon Experiment (EMEX)

During December 1987 to February 1988, the combined Australian Monsoon Experiment (AMEX) / U.S. Equatorial Monsoon Experiment (EMEX) was conducted over the tropical ocean north of Darwin, Australia. Funding for EMEX was provided by NSF and NOAA. It was the first atmospheric process study conducted under TOGA. The experiment investigated the oceanic mesoscale convective systems in the monsoon flow (Webster and Houze 1991). EMEX was designed to test the hypothesis that the net diabatic heating that results from equatorial convection is strongest in the upper troposphere. The net diabatic heating includes offsetting cooling from melting and evaporation of precipitation in and below stratiform cloud regions. The determination of the net heating profile resulting from convection in the equatorial region, and the factors that control it, is critical for predicting the scale of the dynamic response to convective episodes. The heating profile is necessary to understand the tropical forcing of the extratropics, one of the objectives of TOGA. AMEX/EMEX was designed to document, as thoroughly and directly as possible, the vertical profiles of vertical velocity, diabatic heating, and other structures of mesoscale tropical convective-cloud systems ("cloud clusters") over the ocean near the equator. The physical mechanisms responsible for the convective and stratiform components of the observed cloud systems were examined. Airborne and surface Doppler radars, along with other instruments, measured the horizontal and vertical air motions in ten major cloud systems (Holland et al. 1986).

The heating profiles observed during AMEX/EMEX had maxima that were stronger and higher than those observed during the GARP (Global Atmospheric Research Program) Atlantic Tropical Experiment (GATE) (see Figure 3 of Webster and Houze 1991). This finding has important implications for understanding the forcing of the large-scale circulation by convective heating. The information obtained during AMEX/EMEX helped guide the development of TOGA COARE.

Line Islands Array (LIA)

In the early 1980s, various oceanic, equatorially trapped, baroclinic-wave modes that exist in theory were the subject of wide debate. Their actual existence and their hypothesized role in the development of anomalies on interannual time scales of sea surface temperature were questioned. Oceanic synoptic-scale (20–80 day periods) oscillations of sea level, temperature, and currents in the central equatorial Pacific were observed to appear intermittently and with different meridional structures (Legeckis 1977, Hansen and Paul 1984, Wyrtki 1978, Mitchum and Lukas 1987). Synoptic oscillations include variability with time scales of a few days to a few months, as well as phenomena like tropical instability waves[*] and Kelvin-wave pulses. The role of these oceanic synoptic waves in the heat and momentum budgets near the equator was unknown, but thought to be significant. The 1985–1989 Line Islands Array (LIA) experiment in the central equatorial Pacific was deployed to observe synoptic oscillations of sea level in relation to seasonal and interannual variations. Pronounced seasonal and interannual modulation of these synoptic oscillations appeared to be directly forced by both varying wind stresses, and by the strong and variable zonal near-equatorial flows.

The scientific goals of the LIA project were to resolve the meridional sea-level structures associated with these oscillations, correlate these structures with the mean currents, and define the relationship between the growth and decay of the synoptic oscillations and the variation of the equatorial currents. The practical objective of the LIA was to make time series of sea-level observations with an array of instruments between 10°N and 9°S over a five-year period. The LIA was composed of tide gauges and shallow pressure gauges at island sites, and inverted echosounders at deep ocean sites. Five research expeditions were conducted to maintain the array, and to provide shipboard measurements

---

[*] Tropical instability waves are oceanic disturbances found at low-latitude current boundaries, such as between the westward South Equatorial Current and the eastward Equatorial Countercurrent. They are thought to be related to shear-flow instabilities, but may have more complex origin.

of thermohaline and velocity structure. The LIA was funded by NSF and NOAA as a contribution to TOGA.

The LIA captured the full range of interannual variability, including one ENSO warm event and one cold event (Chiswell *et al.* 1995), supporting the overall objectives of TOGA. Interannual and interseasonal variability of dynamic height was dominated by first-baroclinic-mode Kelvin waves. Intraseasonal variability was dominated by an asymmetric mode with maximum energy to the north of the equator. Long records of sea level in the tropical Pacific were analyzed to provide a climatological perspective on sea-level variability with periods between 2 days and 90 days as a function of latitude. The analysis showed pronounced changes with latitude and differences between ENSO and non-ENSO periods (Mitchum and Lukas 1987). The LIA observations have been used to refine the meridional resolution of these results (Donahue *et al.* 1994, Chiswell *et al.* 1995).

Observations from the LIA revealed fluctuations of sea-level, coherent over a wide range of latitudes, with frequencies in the synoptic band (Mitchum and Lukas 1987, Donahue *et al.* 1994, Chiswell *et al.* 1995). The variability in the synoptic band was modulated on longer time scales by ENSO. The 1986–87 ENSO produced changes in the meridional and vertical shears of the zonal equatorial currents, and variations in the Madden-Julian Oscillation (McPhaden and Taft 1988, Enfield 1987). The mean meridional structure of the variability determined from the LIA agreed very well with Geosat altimeter observations and with ocean model simulations (Metzger *et al.* 1992, Donohue *et al.* 1994). Subsequent investigation of the variability in the synoptic spectral band continued during TOGA COARE. Of particular interest, for example, is whether or not there is a downstream rectification of sea-surface-temperature variations associated with synoptic disturbances traveling from the western equatorial Pacific to the central and eastern equatorial Pacific, causing seasonal and longer variations in the central and eastern region.

Tropic Heat

The Tropic Heat program examined the processes that contribute to the maintenance and evolution of the cold-water tongue in the eastern and central equatorial Pacific. It was initiated prior to TOGA; the initial intensive field experiments took place during the boreal fall of 1984. The program continued through a second set of intensive field experiments during boreal spring of 1987. The original objectives were to:
    1. relate the ocean–atmosphere heat flux to oceanic heat storage,
    2. measure horizontal advection of heat in the ocean,
    3. learn how best to measure turbulence at the equator, and

4. develop and understand the modeling of the ocean thermodynamic structure.

The location of the field programs was the region bounded by 5°N, 10°S, 150°W, and 110°W. This region includes most of the cold tongue of the eastern Pacific. The most intensive measurements in 1984 were made at the crossing of the equator and 140°W, near the already-in-place long-term NOAA/EPOCS mooring. Turbulence and fine-structure measurements made at that location were the first systematic and extensive observations of turbulence (sustained for 12 days) taken anywhere in the upper ocean, let alone at the equator. Detailed shipboard measurements of air–sea fluxes were compared to satellite observations, providing early guidance for the TOGA Heat Flux project, as well as for TIWE (Tropical Instability Wave Experiment) and COARE.

Previous notions of a steady-state shear-driven turbulence were demolished by the results of Tropic Heat (Moum *et al.* 1989). It was discovered that solar heating suppressed turbulence in the surface layer, while nighttime turbulence levels extended well below the surface layer into the stratified fluid. Extended analyses demonstrated that the upper 25 meters of the ocean, over broad areas of the equatorial Pacific, loses significantly more heat during night times than during day times.

Tropic Heat II

The focus of observations for the second field program of Tropic Heat, in 1987, was the cycle of deep mixing. Links were observed between the cycle of deep mixing and both internal waves and/or shear instabilities above the core of the Equatorial Undercurrent (EUC). These observations prompted many modeling efforts to explain the physical sequence of events. The significance of internal waves in zonal momentum transport has been elucidated but not yet well quantified. While the 1984 Tropic Heat experiment suggested that internal waves provided a large source of momentum transport, this result was not confirmed by the 1987 measurements. Whether this difference should be attributed to the different seasons of the two experiments, the presence of an El Niño in 1987 but not in 1984, a strong Equatorial Undercurrent in 1984 compared to a weak one in 1987, or simply limited (however extensive) measurement periods, is not known.

One of the original goals of the Tropic Heat experiments was to parameterize mixing, in terms of larger-scale flow indices, for use in models. This goal has proven elusive, as the physics of small-scale processes in the flow field has been found to be quite complex. COARE analyses continue to pursue this objective.

Tropical Instability Wave Experiment (TIWE)

The primary objective of the Tropical Instability Wave Experiment (TIWE) was to describe the space/time evolution of tropical instability waves, with emphasis on divergence, vorticity, and energy propagation. Other objectives were to provide detailed estimates of the Reynolds fluxes associated with these waves and of the role of these fluxes in the heat and momentum balances; to determine the eddy energy balances; to estimate the cross-equatorial Reynolds heat flux attributable to the waves and its divergence; and to estimate the three-dimensional advection of heat associated with the waves. The field campaign was conducted during 1990 and 1991 near 0° latitude, 140°W, in conjunction with the NOAA/EPOCS study of the North Equatorial Countercurrent. Funding for TIWE was provided by NSF.

TIWE consisted of a moored-buoy array along 140°W, where the instability waves are energetic (Halpern *et al.* 1988), together with satellite-tracked drifting buoys and ship surveys. Special attention was devoted to the cusp-like wave structures characteristic of the instability waves evident in fields of sea surface temperature. Conditional sampling of the thermohaline and velocity structures associated with fronts in the upper ocean was assisted by AVHRR images received directly on board one of the research vessels (Flament *et al.* 1996). Drifting buoys revealed the complicated circulations associated with the instability waves, especially the asymmetry of their occurrence with respect to the equator, as well as temperature changes of parcels of water as they move.

North Equatorial Countercurrent (NECC) Study

The North Equatorial Countercurrent (NECC) Study, part of EPOCS, examined the role of the NECC in the heat balance of the eastern equatorial Pacific. The field campaign, which occurred during 1988–1991 near 140°W, was coordinated with TIWE. A moored-buoy array centered at 7°N, 140°W was deployed. The NECC Study provided the first observations using direct current measurements of the interannual variability of the NECC. Other *in situ* measurements came from drifting buoys and ship surveys of upper-ocean thermohaline and current structures. The objectives of the study were to document sea surface temperature, wind, and upper-ocean thermal and flow fields in the NECC; to understand the dynamics of NECC variability and the oceanic response to wind forcing in the vicinity of the Inter-Tropical Convergence Zone (ITCZ); and to provide, within the framework of the TAO array, additional data sets for initializing and validating operational model-based analysis systems. NOAA supported the EPOCS NECC Study.

The annual cycle and a 30-day oscillation in winds and upper-ocean current and temperature in the NECC along 140°W were observed (McPhaden and Hayes 1991). Instability waves with a 30-day period were strongest during July–February, and stronger in 1988 and 1989 (coincident with an ENSO cold phase) than in 1990 and 1991. Variations of sea surface temperature with a 30-day period were greatest near 2°N, where there is a sharp meridional temperature front. However, below the surface, variations in temperature with a 30-day period were greatest around 5°N, associated with vertical and meridional displacements of the sharp thermocline (McPhaden 1996). The latitudinal location of this subsurface maximum in temperature variance coincided roughly with a region of high meridional shear on the southern flank of the NECC, near its boundary with the South Equatorial Current. The energetics of the tropical instability waves, as well as their seasonal and interannual modulation by the large-scale wind-driven circulation, are still to be investigated.

Western Equatorial Pacific Ocean Circulation Study (WEPOCS)

The ocean circulation in the western equatorial Pacific, near the maritime-continent archipelago, was poorly known at the start of the TOGA Program. To improve the description of ocean circulation in the warm pool, the Western Equatorial Pacific Ocean Circulation Study (WEPOCS) was organized as a joint field campaign by the United States and Australia. Several ship surveys making thermohaline and current-profile measurements were conducted between 1985 and 1990. Special efforts were made on some expeditions to estimate air–sea heat fluxes. The sea-level measuring array in the region was enhanced and current-meter moorings were deployed. WEPOCS intended to examine the effect of the monsoon on the upper-ocean circulation in the region north and east of Papua New Guinea, determine the source waters of the EUC, describe the confluence of northern and southern waters in the area that opens into the Indonesian Seas, describe the system of low-latitude western boundary currents, and describe the related deep and intermediate ocean. U.S. participation in WEPOCS was supported by NSF.

An hypothesis that the origin of the EUC is associated with a western boundary current along Papua New Guinea was confirmed by Tsuchiya et al. (1989). The New Guinea Coastal Undercurrent, which supplies approximately two-thirds of the transport of the EUC at its origin, was discovered during WEPOCS (Lindstrom et al. 1987). The other one-third of the EUC transport is produced by the Mindanao Current, which is subject to strong interannual variability (Lukas 1988) around an annual-mean upper-ocean transport of about 26 Sv (Lukas et al. 1991). Both western boundary currents participate in the

Indonesian Throughflow and the equatorial circulation across the Pacific basin (Fine et al. 1994). We have yet to determine how interannual variations of low-latitude western boundary currents influence the warm pool of the western equatorial Pacific.

Coastally trapped Kelvin waves observed using sea-level measurements propagate towards the equator along the coast of Papua New Guinea. The annual cycle of sea level near 7°N, which is strongly modulated by ENSO, resembles a Rossby wave that propagates westward in phase with zonal wind variations, growing in amplitude to the west (Mitchum and Lukas 1990).

Lukas and Lindstrom (1991) discovered the important role of salinity in determining the mixed-layer depth in the warm pool, with salinity changes driven by the heavy rainfall associated with the ascending branch of the Walker circulation. Subduction of warm salty water from the central equatorial Pacific below the less saline water of the warm pool, along with vertical mixing, provides the time-averaged balance to the freshwater flux (Shinoda 1993, Shinoda and Lukas 1995). Questions raised by WEPOCS observations of the upper ocean and measurements of air-sea fluxes were instrumental in the development of TOGA COARE.

Western Tropical Atlantic Experiment (WESTRAX)

The Western Tropical Atlantic Experiment (WESTRAX) was conducted during 1990 to 1993 because of its relevance to TOGA COARE (Brown et al. 1992). Thermohaline and current-velocity-profile measurements were made during ship surveys, current-meter moorings and an echo-sounder array were deployed, and surface drifters and subsurface floats were deployed. The primary objective of WESTRAX was to investigate the mechanisms driving cross-equatorial and cross-gyre fluxes of heat, salt, and momentum. Earlier studies of the western tropical Atlantic region suggested that western boundary currents were continuous from the equator to the Caribbean Sea during boreal spring and were primarily responsible for the fluxes. WESTRAX was supported by NSF and NOAA in the U.S., as well as by France and Germany.

WESTRAX identified several new mechanisms for cross-equatorial and cross-gyre exchanges of mass along the western boundary of the Atlantic Ocean. Johns et al. (1990), using Coastal Zone Color Scanner (CZCS) satellite images of phytoplankton-pigment concentrations and current-meter data, described eddies that separate from the North Brazil Current retroflection and then propagate northwestward along the boundary. These eddies have also been observed in satellite altimetry data (Didden and Schott 1993) and subsurface float data (Richardson and Schmitz 1993, Johns et al. 1990). Didden and Schott (1993) estimated an annual-mean volume transport of 3 to 4 Sv associated with these eddies. Mayer and Weisberg (1993), using climatological data, inferred that cross-gyre transports might occur as a rectification of the annual cycle. Near-surface flow crosses the equator at the western boundary, retroflects into the North Equatorial Countercurrent during boreal summer, and then is swept northward into the subtropical gyre by Ekman transport in boreal winter. About 12 Sv may be accounted for by this mechanism. Also, equatorward cross-gyre exchange was observed. According to Wilson et al. (1994), a portion of the North Equatorial Current returns eastward in both the North Equatorial Countercurrent and the sub-thermocline North Equatorial Undercurrent. The North Equatorial Countercurrent and the North Equatorial Undercurrent also transport water from the North Brazil Current, which was retroflected from the boundary. Northern Hemisphere and Southern Hemisphere water masses are mixed along the western boundary between the North Brazil Current and the North Equatorial Current. An intense subsurface countercurrent extends from the Caribbean to either the North Equatorial Countercurrent or the Equatorial Undercurrent (Molinari and Johns 1994, Johns et al. 1990, Wilson et al. 1994, Colin and Bourles 1994, Schott and Boning 1991). This subsurface current may also contribute to the blending of water masses from the southern and northern hemispheres.

Few of the available data or recent modeling results indicate a continuous northward flow along the western boundary of the Atlantic from the equator to the Caribbean during any season. A monitoring program has been established in the passages of the southern Lesser Antilles to determine whether a significant amount of water from the Southern Hemisphere crosses the equator and enters the tropical–subtropical gyre. Additional modeling and data analyses are being conducted to quantify the roles of continuous boundary currents, eddies, and Ekman fluxes in interhemispheric exchanges of mass and heat.

Equatorial Pacific Experiment (EqPac)

EqPac was a process study of the U.S. Joint Global Ocean Flux Study (JGOFS; Murray et al. 1992), organized in conjunction with TOGA. The purpose of EqPac was to examine the three-dimensional physical and biological processes in the equatorial Pacific cold tongue that determine carbon cycling, the effects of nutrients (including iron) on the biota, and the effects of both on the carbon fluxes. It was conducted in two phases, February–May 1992, when the water was relatively warm (see Figure 9), and August–October 1992, when the eastern Pacific had reverted to relatively normal cold-tongue conditions. The study involved extensive collaboration between NSF- and NOAA-supported components. NSF sponsored process-oriented cruises, mostly on a meridional slice at 140°W from 10°S to 10°N. NOAA supported a semi-synoptic survey consisting of four meridional sections from 95°W and 170°W, providing the context in which the 140°W sections were imbedded. The experiment also included overflights supported by NASA and an iron-fertilization experiment supported by ONR.

The EqPac study found that physical factors in the ocean (thermocline variations, tropical instability waves, and the warm and cold phases of ENSO) are the main factors controlling chemical and biological variability (Murray *et al.* 1994). It was verified that the release of carbon dioxide to the atmosphere varies with the phases of ENSO—smaller during warm conditions (when the thermocline is relatively deep and the carbon-rich waters communicate least effectively with the surface) and larger during cold conditions. Macro-nutrients (e.g., nitrate) were also higher during the cold period, but this elevation contributed little to the chemical and biological differences during the two periods. The particulate export of carbon from the euphotic zone to the rest of the ocean was generally lower in the equatorial Pacific than one would expect from the general level of surface nutrients. Dissolved organic carbon appears to be a significant part of the export of carbon. The relatively low values of primary and export production during the EqPac period, and the differences between the warm and cold periods, were attributed to controls by the composition and size structure of the biological community, possibly influenced by the availability of iron, rather than by the availability of macro-nutrients.

MODELING ENSO

The major modeling advance of the TOGA period was the successful simulation of the ENSO cycle using coupled models of the atmosphere and ocean for the region of the tropical Pacific. Great strides were made in understanding the importance of wind stress from the atmosphere for determining sea surface temperature, and also in understanding the importance of sea surface temperature in forcing atmospheric conditions.

The TOGA decade saw many advances in modeling the ocean and atmosphere in and over the equatorial Pacific. The truly crucial advance was the advent of coupled atmosphere–ocean models for ENSO simulations. Necessary for these coupled models was the development of the ability of ocean models to accurately simulate sea surface temperature in response to the observed winds. While coupled atmosphere–ocean models had existed for 30 years before the start of TOGA, they lacked the near-equatorial spatial resolution needed to represent the waves and upwelling processes essential for the simulation of

ENSO. This section leads to a discussion of coupled models, after first examining the component ocean and atmosphere models, but it should not be construed as a complete review of modeling for the TOGA Program.

Ocean Models

Two major advances in tropical ocean modeling during the TOGA decade were the simulation of sea surface temperature in a variety of ocean models, and the development of an operational hindcast*-analysis system for the tropical Pacific Ocean.

The pre-TOGA years saw the maturation of the linear theory of equatorial ocean waves—their excitation by wind stresses, their propagation eastward as Kelvin modes and westward as Rossby modes, and their effect on vertical displacements of the thermocline. Back then, multilevel primitive-equation models of the equatorial ocean were successful in simulating the thermal structure of the upper equatorial ocean, its current system, and its thermal and momentum responses to changes in the winds (see, e.g., Philander 1981; Philander and Pacanowski 1981a, b). However, sea surface temperature was not routinely simulated. While it was known which processes affected sea surface temperature and its variability (see Sarachik 1985 for a review), the upper-ocean mixing processes were inadequately represented for accurate simulation of sea surface temperature and its variability.

In order to simulate variations of sea surface temperature with a dynamical model, heat and momentum fluxes at the surface must be accurate, upwelling and upper-ocean thermal structure must be correct, and upper-ocean mixing must be well parameterized. A parameterization of vertical mixing based upon the Richardson number (Pacanowski and Philander 1981) allowed the simulated values to be large near the surface of the ocean and very small in the interior, both conditions more realistic than in previous simulations. This representation of upper-ocean mixing then allowed the first reasonable simulation of sea surface temperature anomalies in the context of modeling ENSO (Philander and Seigel 1985).

Parallel to these modeling developments, a simpler approach was pursued by Zebiak (1984). He used a reduced-gravity model with an embedded fixed-depth mixed layer. The mean annual cycle in the ocean was specified and only the anomalies of sea surface temperature were calculated. Because the temperature of the mixed layer was calculated explicitly, the thermal equation for the mixed layer determined the anomalies of sea surface temperature. The temperature of the water entrained at the bottom of the mixed layer was parameterized

---

* A hindcast, sometimes called a "retrospective forecast", is a forward-running simulation using a forecast model started with initial conditions based on an actual past event.

in terms of the thermocline quantities. The Zebiak model proved capable of simulating the anomalies of sea surface temperature characteristic of ENSO.

Heat fluxes at the ocean surface play a central role in controlling the variability of sea surface temperature, especially variability related to the annual cycle. Models with simple parameterizations (see, e.g., Seager *et al.* 1988) have been able to simulate the annual cycle and the full range of variability in the tropical regions to within a degree or two (Seager 1989). For interannual variations, advection—especially upwelling—seems to be the crucial process controlling sea surface temperature in this class of model.

The development of realistic ocean general-circulation models, at the same time that observations of the thermal structure of the upper Pacific became available in real time, encouraged the creation of the first operational ocean-analysis system (Leetmaa and Ji 1989). The data were quality controlled and inserted into a model based on the general-circulation model pioneered at NOAA's Geophysical Fluid Dynamics Laboratory. The hindcast and analysis system interpolates the observations onto a regular grid, thereby creating ocean-wide thermal and flow fields. This operational ocean model became the ocean component of the coupled prediction model at NMC, and its initial state was the initial state for coupled predictions on seasonal-to-interannual time scales.

Atmospheric Models

In order to ensure correct coupling to ocean models, atmospheric models must produce the correct surface fluxes of heat and momentum in response to a specified sea surface temperature. Until the early 1980s, the surface fluxes in atmospheric models were hardly examined; the middle and upper tropospheric fields were considered the *sine qua non* of atmospheric modeling. The need for modeling atmosphere–ocean interactions required improvements in four areas: cumulus convection parameterization, boundary layers in both fluids, and interactions between radiation and clouds.

Cumulus clouds must be parameterized correctly in order to simulate rain in the correct amounts, at the correct places, with the correct cirrus outflows. Both rain and cirrus outflow are important for driving atmospheric circulations. The cirrus outflow helps determine the amount of radiation reaching the surface and hence affects the sea surface temperature. Boundary-layer plumes and shallow clouds mix momentum down from the free atmosphere (say, at 800 mb) to the surface, helping to determine the surface winds. Plumes and clouds also mix moisture from the surface to the top of the boundary layer, helping to determine the amount of moisture available for convergence into precipitating cumulonimbus clouds. Boundary layers in the tropics can be trade-cumulus layers, stratus-topped layers, or layers making the transition between the two.

Stratus-topped layers, which exist mostly over the cooler waters, intercept the incoming solar flux, thereby helping to maintain the relatively cold sea surface temperatures. The interactions of radiation with clouds and with aerosols are crucial ingredients in the correct simulation of the temperature structure of the atmosphere and in the correct simulation of the radiative fluxes reaching the surface.

The simplest models of the tropical atmosphere are based on the work of Gill (1980, 1982). In these two-layer thermal models, the surface winds are determined by the low-level convergence and upper-level divergence required by the thermal forcing. Improvements to these models include convergence feedback (Zebiak 1986) and the combining of the Gill-type model with explicit boundary layers (Wang and Li 1993). The coupling of such models to simple ocean models forms the basis of "intermediate modeling".

The next stages in complexity approach the complexity of full general-circulation models. The insertion of complex boundary layers, both trade-cumulus and stratus topped, are major undertakings and have not yet been fully implemented in general-circulation models, although it is known that their absence severely limits the accuracy of coupled models.

Coupled Models

The major modeling advances made during the TOGA period involved the successful simulation of the ENSO cycle by means of coupled atmosphere–ocean models of the tropical Pacific. The first successful coupled model of ENSO consisted of a Gill-type model (Gill 1980) of the atmosphere, with improved moisture convergence (Zebiak 1986), coupled to a reduced-gravity ocean model with an embedded surface mixed layer (Zebiak and Cane 1987). A novel feature of the Zebiak and Cane model was a fixed-depth frictional surface layer embedded in the uppermost dynamic layer of the modeled ocean, allowing a simple representation of the near-surface intensification of wind-driven currents in the real ocean. A complete thermodynamic equation was used for this surface layer, including all advection and upwelling terms. This model proved successful at simulating the time development of sea-surface-temperature anomalies during an ENSO cycle, at generating the correct time scale of recurrence of warm and cold events, and in producing reasonable decadal variability of the warm and cold phases of ENSO. This coupled model also predicted the 1987 warm phase of ENSO a full year in advance, and thereby provided the first successful short-term climate prediction based on a coupled model.

The success of the Zebiak and Cane (1987) models provided the encouragement to couple more realistic and complex atmosphere and ocean general-circulation models. The complexity and difficulty of the task proved greater than foreseen (Neelin *et al.* 1992). It was well understood that the ocean component had to produce the correct sea surface temperature in response to the fluxes of heat and momentum from the atmosphere. It was also understood that the modeled atmosphere had to produce the correct surface fluxes in response to the sea surface temperature. What was not appreciated is how sensitive each model is to errors in the other.

The original attempts to couple the atmosphere to the ocean ran into problems; the final state turned out to be far from a realistic one (see, e.g., Manabe and Stouffer 1988). In order to restore a more realistic climate state, the artificial expedient of "flux corrections" was adopted, viz., the arbitrary imposition of surface fluxes designed to force the modeled state to a realistic one (Saucen *et al.* 1988). These flux corrections have proven detrimental to simulations of climate change because they always tend to bring the climate state back to the original one, regardless of the state to which the changes in the forcing should impel the model. Improved coupled models have produced realistic interannual variability, representative of ENSO, without the need for flux corrections (Philander *et al.* 1992, Nagai *et al.* 1992), but only in the absence of the annual forcing of insolation.

The annual cycle is subtle and difficult to simulate. It involves processes in the atmosphere and ocean different from those governing interannual variability (Köberle and Philander 1994). In particular, heat fluxes play a more important role in the annual cycle than the interannual cycle. For example, stratus clouds seem to be crucial for modulating the heat fluxes needed to produce the annual cycle, especially in the coastal regions of the equatorial and South Pacific (Klein and Hartmann 1993). In the cold-tongue regions, stratus clouds influence the far-eastern Pacific (say within 1000 km of the coast), where the variations of the heat fluxes through the ocean surface are in phase with the variations of sea surface temperature. However, stratus clouds appear to be unimportant for modulating interannual variations of the cold tongue in regions away from the coast, where the variations of heat fluxes and sea surface temperature are out of phase. The unrealistic warmth of *all* existing coupled general-circulation models at the far eastern part of the cold tongue has been attributed to the inability of the models to successfully simulate the stratus clouds and their effects (Mechoso *et al.* 1995).

At this writing, coupled models are just beginning to produce combined annual and interannual sea-surface-temperature variability without the artificial expedient of flux corrections (Latif *et al.* 1993, Schneider and Kinter 1994, Robertson *et al.* 1995a), although the modeled variability is not yet totally realistic (Mechoso *et al.* 1995). The challenge remains for coupled models to simulate successfully both annual and interannual variability in the tropics.

PREDICTION

All objectives of the TOGA program were related to improving prediction, or quantifying predictability, of short-term climate variations. Progress towards achieving these objectives ranks among the greatest successes of TOGA. Not only was the basis of climate predictability established through ENSO theories and models, but by the end of TOGA, regular and systematic experimental ENSO predictions had been underway for several years.

It is generally thought that the intrinsic variability of the atmosphere creates a barrier to predictions beyond approximately two weeks (Lorenz 1965, 1982; Charney et al. 1966). However, the slowly evolving nature of the ocean, together with the strong coupling between the ocean and atmosphere, allows that barrier to be avoided. Charney and Shukla (1981) pointed out that the muted high-frequency atmospheric variability of the tropics made the atmosphere there highly dependent on its boundary conditions, and therefore much more predictable on time scales associated with the slow variation of those boundary conditions. Prediction of sea surface temperature, a boundary condition for the atmosphere, allows the forecast of concomitant atmospheric statistics.

Predictions of eastern tropical Pacific sea surface temperature can be made a season to a year or more in advance (e.g., Latif et al. 1994). While weather cannot be predicted further ahead than a week or two, seasonal climatic variables for particular regions of the world directly affected by ENSO *can* be predicted. The TOGA period has seen this concept developed and demonstrated in practice. The TOGA Program fostered development of both statistical and dynamical prediction models, established a coordinated prediction program that is the prototype of a multinational climate-prediction enterprise, and created operational climate predictions. Prediction products are routinely disseminated and used worldwide.

Many diverse definitions of forecast skill have been used to evaluate weather predictions. At this early stage in the development of climate forecasts, only simple measures of skill are in regular use. The most common measures are (1) the correlation of a forecast with the NINO3 index and (2) the average of the root-mean-square error over the length of a forecast. These two measures give a simple indication of how well phase and amplitude have been predicted over the length of a record. More sophisticated measures compare the success of a forecast to the success of persistence and/or climatology.

It should be noted that there is predictability in the sea-surface-temperature fields simply because of their persistence. The autocorrelation of the index called NINO3 (the anomaly of sea surface temperature averaged over the region between 5°N and 5°S, 150°W and 90°W), for example, shows a 0.5 autocorrelation with a lag of 6 months. Clearly, the correlation of a prediction of NINO3 with observations must beat this value in order to demonstrate useful skill. With a lag of 3 months, the autocorrelation of NINO3 is about 0.7. Skill in predicting sea surface temperature, as measured by the correlation of observed and predicted time series, is expected to have its greatest value for lead times exceeding 6 months, when the value of persistence forecasts is low.

Inoue and O'Brien (1984) provided the first ENSO predictions using a numerical model. In their prediction scheme, a dynamic ocean model forecast the evolution of the thermocline depth in the eastern Pacific, assuming that surface wind-stress anomalies remained fixed at their initial values. The model clearly demonstrated a predictive potential at lead times of several months, even though sea-surface-temperature and atmospheric fields were not explicitly forecast.

Prediction schemes for ENSO based on statistical models were introduced by Graham et al. (1987a, b). In these schemes, sea-level pressure or tropical Pacific winds are used to predict sea surface temperature. A significant issue in the construction of such models is the appearance of artificial skill, i.e., the tendency of a statistical scheme with many degrees of freedom to produce results which are arbitrarily good in hindcast mode (reproducing the data used to construct the model), but poor in true forecast mode. Graham et al. minimized artificial skill by reducing the degrees of freedom in the model through the use of empirical orthogonal functions (EOFs). Others have used principal-oscillation patterns (Xu and von Storch 1990; Penland and Magorian 1993). The results of these schemes are clearly superior to persistence forecasts, providing correlations of about 0.5 with lead times of 9 months for predictions of sea surface temperature in the eastern Pacific (Latif et al. 1994, Barnston and Ropelewski 1992). Prediction correlations at this level using statistical models demonstrate clearly the large-scale, low-frequency nature of ENSO, as well as a significant degree of linearity in its evolution over time scales of several seasons.

By the close of the TOGA Program, monthly statistical forecasts were being made by the Climate Analysis Center for 3-month-mean sea surface temperature in several regions of the tropical Pacific and Indian Oceans (Graham et al. 1987a, b; Barnston and Ropelewski 1992). Forecasts were made with lead times up to 12 months. Results indicated that the 170°W–120°W region of the equatorial Pacific is the most predictable. For forecasts with a lead of 6 months, the anomaly correlation score for the last decade over this region is 0.7. Skill is greatest for boreal winter forecasts made at the end of the summer. Forecasts made during winter for the following summer exhibit considerably less skill.

Statistical forecasts outperform persistence in the equatorial Pacific under most conditions. Similar statistical techniques have also been implemented to forecast North American air temperature and precipitation (Barnston et al. 1994), but with forecast correlations for seven-month lead times not exceeding 0.3.

The first coupled atmosphere–ocean model applied to ENSO prediction was that of Zebiak and Cane (1987). Only departures from monthly-mean atmospheric and oceanic climatologies were explicitly calculated. As the climatology was *specified* from observations, the mean state was guaranteed to be realistic. Atmospheric dynamics were approximated by a steady, linear, "shallow water"* system, following the work of Gill (1980) and later modified by Zebiak (1982, 1986) for use in modeling ENSO. Oceanic dynamics were also approximated by linear, shallow-water equations, following the work by Busalacchi and O'Brien (1980, 1981). The Zebiak and Cane model simulated a realistic ENSO. Analyses by Zebiak and Cane, Battisti (1988), Suarez and Schopf (1988), and Schopf and Suarez (1988) highlighted the crucial role of ocean dynamics in supplying a delayed negative feedback that sustained the oscillations in the models. The "delayed action oscillator" theory of ENSO (Battisti and Hirst 1989; Wakata and Sarachik 1991a; Jin and Neelin 1993a, b; Neelin and Jin 1993) is now the most widely accepted explanation of ENSO (see Battisti and Sarachik 1995 for more details).

Regular predictions of ENSO on an experimental basis, using the Zebiak and Cane model, began in 1985 (Cane et al. 1986). Figure 10 shows the evolution of the correlation skill for a set of hindcasts of the NINO3 index for El Niño events of 1976 and 1982–1983, and the nonevent years of 1977–1979. The forecasts of these different periods were all largely successful at lead times of a year or more. The overall correlation skill (i.e., the correlation between the predicted and observed index) for NINO3 forecasts during this period was above 0.6 for lead times up to 11 months. Such success is particularly remarkable in view of the highly simplified initialization procedure that was used. Initial conditions of the ocean for each forecast were obtained by forcing the ocean model with analyzed surface winds based on data from VOS (Goldenberg and O'Brien 1981), often marginal in quality because of poor coverage. Initial conditions of the atmosphere were derived from the ocean-model simulation of sea surface temperature. Thus, no observations of sea surface temperature or subsurface temperatures were utilized. That such a simple scheme could allow skillful predictions strongly reinforces several prevailing ideas about ENSO: it is a large-scale, low-frequency signal; the important "memory" of the coupled system resides in the pattern of upper-ocean heat content; and, finally, the heat-content variability is largely controlled through surface forcing by wind stress,

---

* The so-called "shallow water" equations are approximations used for many fluid systems, including the atmosphere, that have much greater horizontal expanse than depth.

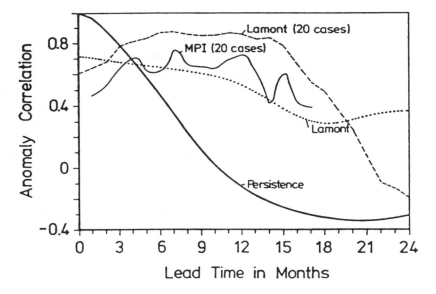

Figure 10. Correlations for predictions of anomalies of equatorial sea surface temperature using fully coupled atmosphere–ocean models. The curve labeled "Lamont" shows results obtained with the coupled model of Zebiak and Cane (1987). The curve labeled MPI shows results obtained with the coupled general-circulation model of Latif *et al.* (1993). The curve labeled "Lamont (20 cases)" shows the anomaly correlations for the Lamont model when applied to the same 20 cases to which the MPI model was applied. Also shown, for reference, is the result of persistence forecasts. (Reprinted with permission from Latif *et al.* 1994, copyright Springer-Verlag.)

in accordance with linear, shallow-water dynamics. The results from coupled models justify an optimistic assessment of the additional predictive potential of more detailed models and assimilation systems that could take advantage of all available observations. We can clearly see from the figure that while the overall skill of the Zebiak and Cane model is useful, there are specific times when the forecast skill is considerably higher than other times. Conversely, predictions made during boreal spring are considerably worse than the mean. The skill of the Zebiak and Cane model has been recently increased by more careful initialization, which not only reduced the nowcast error, but also increased the correlation skill out to more than a year (Chen *et al.* 1995).

The first ENSO forecast using coupled numerical models, made in early 1986 (Cane *et al.* 1986), predicted a warming event for later that year. The forecast for January 1987, made with a lead time of one year, was essentially accurate, although the rapid warming in 1986 occurred about 3 months later than had been predicted. Barnett et al. (1988) compared the predictions for 1986–1988 made using the Zebiak and Cane model with predictions from the statistical forecast model developed by Graham et al. (1987a, b), and found similar performance at lead times of 9 months or less. Evaluated over a longer record, the performances of the dynamical and statistical models are comparable for lead times of 4 months or less, but dynamical predictions are increasingly superior at longer lead times (Latif et al. 1994).

In 1991, after informal discussions with members of the TOGA Panel, the TOGA Project Office began a program to further develop the new field of short-term climate prediction. The program, the TOGA Program on Prediction (T-POP, described in Cane and Sarachik 1991) was designed to be inclusive, supporting all the ongoing research in the United States on short-term climate prediction. Its focus was the development of prediction systems—models, data-assimilation techniques, and data quality control—based primarily on coupled atmosphere–ocean general-circulation models. In addition to prediction research, T-POP concentrated on making experimental predictions and on inter-comparisons of predictions for designated time periods.

Forecasts using coupled models require initialization of only the ocean components of the model because, in most instances, information about the atmospheric initial state is lost within the first two weeks of the model run. The simplified prediction systems (e.g., the one developed by Cane and Zebiak) use the history of wind stress to initialize the state of the ocean. During TOGA, a real-time ocean-analysis system (Ji et al. 1994b) was implemented at NMC. The NMC system provided forecasts of sea surface temperature for the tropical Pacific, as well as forecasts of rainfall and surface temperature around the globe. The skill level of this model for about the first two seasons was higher than that of other systems because an observed ocean thermal state is used for initialization rather than a proxy based on the history of the winds (similar results were obtained by Rosati et al. 1996). Forecast experiments where subsurface data were not used in the assimilation show a loss of skill, measured by correlation coefficients, from about 0.8 to 0.6 at two seasons.

Because much of the global atmospheric variability depends on variations of tropical sea surface temperature, any model that forecasts sea surface temperature can be used with any atmospheric general-circulation model to forecast atmospheric variations around the globe. This process has several advantages in opening up the field of climate forecasting to groups that do not have the resources to run fully coupled models. Groups that possess only atmospheric models can make seasonal-to-interannual climate forecasts by using one or all of the several, now routinely available, forecasts of sea surface temperature.

The TOGA goal of establishing the existence of seasonal-to-interannual climate predictability has been realized. Actual predictions of those aspects of climate associated with ENSO have become available in increasing numbers from a suite of dynamical and statistical models. Stemming from these achievements has been a growing sense of the opportunity to capitalize more fully on the established predictive potential for societal and economic benefit. While the examples provided in chapter 6 demonstrate actual benefits of climate forecast applications, they point to the even greater potential that would accompany a more organized and focused climate prediction and application activity. It was with this motivation that the concept of an International Research Institute for Climate Prediction (IRICP, see chapter 7) was developed.

As of the end of TOGA, routine predictions of sea surface temperature based on ENSO dynamics were being made using the Zebiak and Cane model, various statistical models, and the NMC coupled model. These predictions were published in the NOAA Climate Diagnostics Bulletin and the NOAA Experimental Long Lead Forecast Bulletin. Additional forecast products were available from several other groups, both national and international. Products consisted of predicted indices (e.g., NINO3), as well as predicted fields, at lead times ranging from 1 to 18 months. The routine dissemination of such products attested to a high level of maturity of ENSO prediction, and signaled the movement into an operational mode in which practical application of climate predictions can begin. However, experience has shown that there can be strong intradecadal variations in the skill of forecasts. During the 1980s, when ENSO was regular and of large amplitude, forecast skill was high. Events during the 1990s have proven more difficult to predict (see section on The Warming During 1990–1994, p. 85). Appendix B lists known prediction products.

TOGA PRODUCTS

As a result of the building of the TOGA Observing System and the development of a predictive capability for aspects of ENSO, a large number of products (both observational and predictive) have become available (see Appendix B for a list). These products encapsulate the variety of efforts and progress of TOGA. They are generally available, they are used and consulted widely, they have many sources (purely observational, model-assimilated, and predictive), and they are available in different media (paper and electronic) through a variety of distribution networks.

These products were motivated by TOGA, a research program, but have transcended it. Normally, when a research program ends, what is left are the data and the resulting knowledge. TOGA has accomplished more than that. The key to understanding this is TOGA's role in the development of short-term climate prediction. Products exist, and will continue to be issued on a regular basis, because predictions will continue to be made as long as they are found to be useful. TOGA has provided products because the community that produces and uses short-term climate prediction demands that the ocean measurements started by TOGA be continued and be made widely available.

An analogy can be made between maintenance of the upper-air observing system in support of weather prediction and maintenance of the ocean observing system in support of climate prediction. The upper-air network has been maintained for about 50 years not so much because it is producing good science (which it is), but rather because it has proven useful to society. We have gained a global view of the atmosphere and its variations because weather prediction is recognized to be valuable. Similarly, the ocean-data products developed in conjunction with ENSO predictions are already used by society and have a recognized value beyond their scientific worth. We expect that our knowledge of the ocean will be expanded because of the maintenance of an ocean observing system in support of climate prediction.

Reanalysis

Reanalysis is a tool that allows consistent data sets to be produced for climate studies. It is needed because existing weather analyses are inadequate for use in climate studies. Climate data sets were needed during TOGA both for diagnostic studies and for evaluation of predictive systems using historical events. For studies of interannual climate variations, a consistent record spanning several decades would be very useful. The long record of weather forecasts and analyses produced by national weather centers would seem to be an ideal data set for climate studies. However, many changes—mostly undocumented—have been made in the various components of the weather forecasting system. Changes in the quality-control protocols, in the data-assimilation models and methods, in the initialization procedures, and in the weather prediction models all contribute to spatial and temporal inhomogeneities in the long-term record of weather analyses.

The only way to "undo" these changes and to produce a spatially and temporally homogenous set of analyses suitable for climate studies is to repeat the entire procedure with the best available weather-data assimilation systems. NMC agreed to reanalyze the entire climate record beginning with 1960, work that it continued as NCEP. In response to a request from T-POP, the first five years to be reanalyzed will be those from 1985–1990, to facilitate the intercomparison of predictions that T-POP has undertaken for a common period. NASA Goddard has also performed a five-year reanalysis, for March 1985 to February 1990, as part of the Earth Observing System (EOS) project. NASA is also preparing a specialized reanalysis of data from the western Pacific in support of COARE. NCEP and the ECMWF have also performed reanalyses of data from the COARE period.

Reanalysis at a single point in time does not solve the problem posed by further additions to the climate record. To have a useful, growing climate record, reanalysis must be done regularly, say every ten years or so. Reanalysis not only allows an orderly growth of the climate record, but also allows data that did not arrive in time for the original analyses (and are therefore normally wasted) to be put to use. Regular reanalysis provides continuing incentives for uncovering and rehabilitating old data.

TOGA CD-ROM Project

The ITPO worked past the end of TOGA to make the TOGA data available on CD-ROM. A wide range of data sets for the ten-year TOGA period, provided by the TOGA data centers and by other data centers in various nations, are freely available to the international research community on CD-ROM. As a first step, data sets for 1985 and 1986 were published (Halpern *et al.* 1990) on one CD-ROM, which has been widely distributed. Subsequently, CD-ROMs with data up to 1990 were prepared at the Jet Propulsion Laboratory (JPL) for distribution in 1994, with further CD-ROMs to follow.

PROBLEMS AND SHORTCOMINGS

Although the U.S. TOGA Program accomplished much, it did encounter problems. Discussion of some of these can also be found in NRC 1992. Whether some aspects of the program should have, or could have, been handled differently is a matter of judgment. On some of the issues mentioned below, no consensus has been reached on whether these were truly shortcomings of the program. Problems involved: the overall scope of the program, the development of observational systems, the timing and relevance of the field programs, and the cooperation among all the participating individuals and organizations. However, resolution of some of these problems within the scope and lifetime of the TOGA Program was one of the strengths of the program.

The original plan from the United States (NRC 1983) for the program that became TOGA concentrated solely on ENSO in the tropical Pacific. The international TOGA plan (WCRP 1985), however, took a more expansive view of the problem and proposed a more general study of seasonal-to-interannual variability throughout tropical oceans and the global atmosphere, with the objectives previously listed (p. 19). These broader objectives were accepted by the U.S. community (NRC 1986). However, these objectives proved overly ambitious. The financial resources from the United States, and the rest of the international research community, for such an ambitious program were not available. Nor was any great effort made to distribute the study and financial responsibilities among the participating nations to cover the broader objectives of the global program. Even if greater financial resources had been available, it is not clear that a sufficient number of trained scientists interested in these problems would have been immediately available. In practice, TOGA resources were focused on studying interannual variability in the tropical Pacific region, falling back to a strategy more closely resembling the 1983 plan. This limited focus may have contributed to the overall success of TOGA. However, even the great attention paid to ENSO left unanswered many questions about the annual cycle in the most studied region of the tropical Pacific.

The original plans for TOGA relied heavily on proposed and planned observations from satellites. Insufficient attention was paid to the political and operational realities of satellite programs, and the timing of when satellite-based observations of the physical quantities most needed for TOGA would really be available and reliable. When the satellite programs on which the original plans relied failed to materialize, the TOGA Program had to develop alternate strategies. While the resulting TOGA Observing System, and its array of moored buoys in the tropical Pacific, was one of the great achievements of the program, much of the system was not available until nearly the end of the program. The usefulness of the system for making forecasts of seasonal-to-interannual climate variations has still not been fully evaluated. The best mix of satellite-based and *in situ* observations for research or operational forecasting is still unknown. Problems also remain as to how to arrange for research observing systems to become part of the operational observing systems. Furthermore, continuity is not assured for observations of the many geophysical quantities needed for studying and predicting seasonal-to-interannual climate variations.

The relevance of some of the field programs of TOGA to the larger program has been the subject of debate. The several field programs discussed earlier in this chapter, especially the large COARE field program, were some of the most expensive parts of TOGA. Though there is little dispute that these programs, especially COARE, were, and will continue to be, valuable for understanding air–sea interaction, some scientists would have preferred that the resources had been allocated to examining processes on the larger spatial scales more clearly relevant to directly attaining the objectives of TOGA (although new funds were found for the COARE field program). COARE was not completed until near the end of TOGA and the data from the field program are still undergoing intense analysis; it therefore did not have a large direct influence on TOGA. Though COARE has already produced much excellent science, the extent to which it will produce knowledge applicable to the basic objectives of TOGA is not yet known.

The TOGA Program required cooperation among scientists from several disciplines, several government agencies, and scientists and organizations from several nations. The required cooperation did not always develop smoothly. Although by the end of TOGA, participating atmospheric and oceanic scientists had learned to work together, and the larger community was viewing the atmosphere and oceans in a more unified way, at the beginning of the program lines of communication were not as open (see p. 106). Data management, though kept relatively inexpensive, relied heavily on the work of the scientists collecting the data, so that not all data were immediately available to the larger community or the operational centers that could use them for real-time model initialization. However, data accessibility improved as the program evolved, to the extent that all data from the TAO array are now freely available in real time. Furthermore, for the results from TOGA to be valuable to a large user community, applications must be developed (see p. 124), but communication and understanding between physical and social scientists is still difficult, and no coherent strategy for identifying and effecting applications has yet arisen.

Interagency cooperation presented difficulties because of the differing objectives and operating styles of the agencies involved. NOAA had a specific operational mission guiding its research. NASA had a research strategy more linked to development of space-based technology than to specific research problems. NSF preferred to respond to the best scientific proposals it received without promising funds for a program. These agencies struggled to coordinate their activities. The problems are structural and will face any future large programs for climate research. The facts that this report concentrates so heavily on the U.S. efforts, that not all financial data from participating U.S. federal agencies are uniform and available for all years of the program, that financial data from other countries are limited, and that some issues remained unresolved through the preparation of this report on the relationship of the advisory and review mechanism (e.g., the TOGA Panel) to the program and the several sponsoring agencies all point to the difficulties of coordinating the large international TOGA Program. However, such problems confront all large cooperative ventures.

4. What We've Learned

TOGA was devoted to a study of ENSO in and over the tropical Pacific, its effects over the tropics and into midlatitudes, and its predictability. During the ten years of TOGA, the deployment of an observational system in the Pacific enabled us to observe the evolution of two warm phases of ENSO (1986–1987 and 1991–1992), one cold phase (1987–1988), and the prolonged warmth in the Pacific lasting from 1990 to 1994. The beginnings of a mathematical theory of the oscillatory aspects of ENSO have been developed. The predictability of sea surface temperature in the eastern Pacific, with lead times of about a year, has been demonstrated. Explorations of the effects of ENSO on the rest of the globe were pursued. The first short-range climate forecast made using coupled dynamical models predicted a year in advance the warming near the end of 1986. The TOGA years have seen major advances for the observational and theoretical understanding of ENSO. A series of regularly appearing data and prediction products capitalize on these advances. The meteorological and oceanographic communities worked together to bring about these accomplishments, reducing the barriers between their disciplines in the process.

We now have instruments to observe the evolution of the coupled atmosphere–ocean system in the tropical Pacific and theories to describe ENSO as a coupled atmosphere–ocean phenomenon. The observational progress made during TOGA has enabled us to begin evaluating theories of ENSO. The most apparently successful theory of ENSO is the delayed oscillator. This theory applies only to the completely regular case, although aspects of the theory apply to the irregular case as well. The underlying cause of the irregularity of ENSO can be narrowed to two possibilities: noise or nonlinearities, perhaps both. These observational and theoretical advances were major scientific lessons of TOGA. The program also provided lessons on how to conduct integrated research on the climate system.

OBSERVATIONS OF ENSO IN THE TROPICAL PACIFIC

Climatologies

ENSO is, among other things, an anomalous warming of the eastern Pacific. In order to define the anomalies, it is crucial to define the background, normal conditions against which the anomalies are measured. While simple in concept, this observational and definitional problem is complicated by the existence of both interannual (and longer) variability and intraseasonal (and shorter) variability. In the implicit definition used throughout this report, the climatology of a quantity is the sequence of monthly averages of that quantity for all the months of the year. Because both warm and cold phases of ENSO contribute to the monthly averages, a record long enough to include the effects of the slow interannual variability must be obtained. If the annual average is not stationary, the climatology will be unstable—i.e., different climatologies will arise from different averaging periods. Attempts to define a climatology by averaging only during "normal" periods—i.e., those without significant warm or cold phases of ENSO—will give an incorrect climatology if ENSO produces rectified effects.

Sea surface temperature is one of the key quantities that change during ENSO, and its climatology is therefore one of the most crucial. To date, climatologies of sea surface temperature have been obtained predominantly from historical records of *in situ* data (e.g., Reynolds 1982, Slutz *et al.* 1985), or by accumulating statistics from operational analyses (e.g., Reynolds and Smith 1994). Climatological winds have usually been obtained from historical records (e.g., Hellerman and Rosenstein 1983, Harrison 1989). The TOGA Observing System is providing large numbers of tropical data, which future climatologies will reflect.

The TOGA Observing System has produced some remarkable results on the climatologies of the oceanic subsurface thermal structure and subsurface circulation. Current-meter moorings have been in place at 110°W since 1980 and at 140°W since 1983. In conjunction with ATLAS moorings, they have yielded a remarkable picture of the behavior of the near-surface circulation (see, e.g., McPhaden and McCarty 1992). At 110°W (see Figure 7, middle), for example, the undercurrent is strongest in boreal spring when the winds are weakest, the sea surface temperature is warmest, and the surface currents have reversed to eastward. The warm sea surface temperature has no thermocline motion associated with it—indeed, the thermocline stays pretty much flat throughout the year. Clearly then, the main processes available to change sea surface temperature are the surface fluxes and surface advection, both affecting (and affected by) the mixing processes that determine the mixed-layer depth.

Models have been used to understand aspects of this evolution of the thermal structure (see, for example, Yin and Sarachik 1993 for a study using a two-dimensional nonlinear undercurrent model, and Philander *et al.* 1987 for a study using a three-dimensional ocean general-circulation model). The interpretation developed from modeling studies is that during boreal spring, when the winds are weak and the upward vertical advection of eastward momentum from the equatorial undercurrent is weakest, the pressure gradient is still strong and the surface currents are accelerated eastward, mostly by the pressure gradient. The surface currents then advect the temperature down gradient and warm the eastern part of the region. Vertical advection of momentum cannot be measured directly, but such advection plays an important role in the maintenance and variation of the undercurrent. As time goes on, and if the TOGA Observing System stays in place, we may expect to have a more complete, basin-wide view of the climatology of subsurface thermal variations.

Evolution of Warm and Cold Events

The canonical ENSO described by Rasmusson and Carpenter (1982)—on the basis of warm phases occurring in 1951, 1953, 1957, 1965, 1972, and 1976—showed a definite westward propagation of anomalies of sea surface temperature at the same time that isotherms of the total sea surface temperature moved eastward. This was possible because the isotherms of the annual cycle had a well defined westward propagation (see Figure 11a), and any anomaly added to the annual cycle would move isotherms farther east.

Descriptions of the propagation characteristics of the 1982–1983 warm phase are not consistent in the literature. However, it appears that the anomalies of sea surface temperature (see Figure 11b) warmed uniformly over large parts of the central Pacific, followed by warming near the coast of South America. It is clear that in some sense the anomalies of sea surface temperature did propagate eastward, but in another sense the warm sea surface temperature developed over a large region of the tropical Pacific without propagation.

Figure 9 shows the anomalies, relative to the climatology of Reynolds and Smith (1994), of sea surface temperature along the equator during the TOGA years. All points of view about the propagation of sea-surface-temperature anomalies can find evidence in this series. The anomalous warming in 1991 seems to have a component that propagated eastward, although in the boreal spring of 1991 warm anomalies appeared simultaneously in the east and around the dateline. The anomalous warmth of early 1993 seems to have set in simultaneously across the entire eastern Pacific, with no obvious phase propagation. The cooling of 1988 appears to have a definite westward propagation, although the zero line of the temperature anomaly shows no phase propagation.

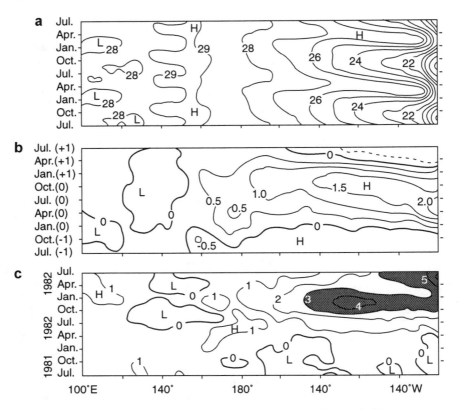

Figure 11. Evolution of sea surface temperature in the tropical Pacific. The sections follow the equator from the western Pacific up to 95°W, then follow the climatological cold axis to the coast of South America, reaching 8°S. Panel (a) shows the mean climatology of the annual cycle, repeated for comparison with the other panels. Panel (b) shows anomalies for a composite El Niño, with the year of the warm peak designated year 0. Panel (c) shows the anomalies for 1981–1983. Note the difference in the contour intervals. (Reprinted with permission from Cane 1983, copyright American Association for the Advancement of Science.)

A more complete picture of the evolving warm and cold phases of ENSO (including subsurface thermal structure) is now available as a guide to modelers. The winds, subsurface thermal structure, sea surface temperature, and sea level are measured directly by the TOGA Observing System, and other quantities such as precipitation and surface fluxes are obtained by other methods. These should become available in future years from the TAO array, which was completed at the end of TOGA.

The Warming During 1990–1994

An increased incidence of warm phases of ENSO since about 1970 is clear in the instrumental record (Smith *et al*. 1994). The warming from 1990 to 1994 is unprecedented in the instrumental record (Trenberth and Hoar 1996). The Southern Oscillation Index (SOI) was negative during those five years, the western Pacific around the date line has been anomalously warm for most of that period, there were two warm phases of ENSO very close to each other (peaking in the boreal springs of 1992 and 1993) and another one well developed at the end of 1994. In addition, there was a persistent "horseshoe-shaped" anomaly of sea surface temperature developing early in 1990 (see Figure 12) extending from 20°S to 20°N in the east Pacific and crossing the equator at the dateline in the west. The predictions of sea surface temperature in the NINO3 region have been poorer during this time, with some notable misses by all dynamical and statistical models.

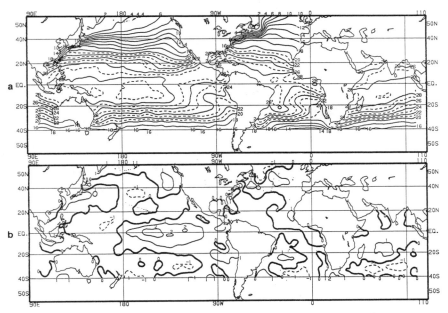

Figure 12. Total field (upper) and anomalies (lower) of sea surface temperature for January 1992. (From Climate Analysis Center 1992.)

Whether the decadal mean temperatures in the eastern Pacific have changed (caused by some possible natural fluctuation or by anthropogenic greenhouse warming) or whether the frequency of ENSO warm events has changed due to some other cause is, as yet, impossible to say. Also unusual in the period 1990–1994 is that the magnitude of the anomalies of sea surface temperature in the equatorial Pacific peaks in the boreal spring, coincident with the normal annual warming, rather than the normal case for ENSO with the magnitude of the warm anomalies peaking in late boreal winter.

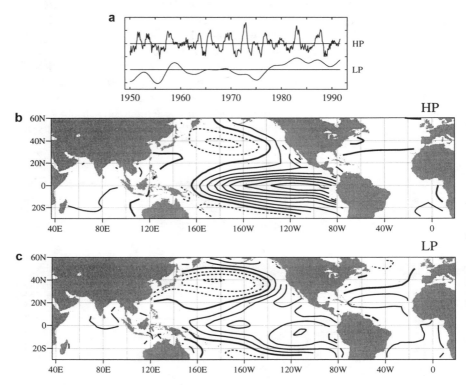

Figure 13. Analyses of interannual and lower-frequency variations of sea surface temperature throughout the entire Pacific. Panel (a) shows the amplitude as a function of time for the leading (normalized) principal components with a high-pass (HP) filter, panel (b), and a low-pass filter (LP), panel (c), divided at 6 years. The tick marks in panel (a) indicate 1.0 standard deviations. The contour intervals in panels (b) and (c) correspond to 0.1 K per standard deviation. Negative contours are dashed; the zero contour is thickened. (From Zhang et al. 1996, reprinted by permission of the American Meteorological Society.)

Since the end of TOGA, the nature of the unusual warming of the early 1990s has gradually become clearer. Figure 13 shows analyses of both interannual and lower-frequency (periods longer than a few years) variations of sea surface temperature throughout the entire Pacific (Zhang et al. 1996; similar results were found earlier by Nitta and Yamada 1989). The pattern of the higher-frequency variability of sea surface temperature is clear—it is the ENSO phenomenon tightly confined to the equator, juxtaposed with a weak (opposite sign) North Pacific covariation. The lower-frequency variability is much less spatially confined to the tropics—it exhibits a horseshoe-shaped pattern similar to the pattern that persisted throughout the early 1990s, as shown in Figure 13. The amplitude function (the lower curve of Figure 13a) shows that the decadal mode was strong during the early 1990s and that it indeed provided a different background state for the more equatorially confined ENSO variations.

We now know that skill for predicting variations of sea surface temperature near the equator varies decadally (see, for example, Balmaseda et al. 1995 and Chen et al. 1995). It seems reasonable to speculate that the varying background state provided by the decadal mode is the cause of the decadal variation of ENSO prediction skill. Examination of the decadal time series in Figure 13 would seem to indicate that the decadal mode has been strong since the 1980s (when predictive skill was high), so that if this speculation is even partly true, other factors currently unknown must be involved. While the decadal mode has been identified in observations (Balmaseda et al. 1995, Chen et al. 1995) and simulated in a coupled general-circulation model (Latif et al. 1996) the mechanism for this mode is still not clear. Latif et al. (1996) argue that this mode is fundamentally a midlatitude mode with a tropical expression; they also provide a mechanism that depends essentially on midlatitude dynamics. Consistent with this view is the observation that the midlatitude expression of the decadal mode is stronger than the tropical expression, in contrast to the interannual mode, where the midlatitude expression is distinctly weaker. Whatever the cause, it becomes important to understand this decadal modulation of ENSO in order to improve prediction skill.

EFFECTS OF ENSO ON THE REST OF THE GLOBE

Tropics

The interactions on interannual time scales of the tropical oceans, communicating through the medium of the atmospheric circulation, were investigated by Latif and Barnett (1995), who conducted a series of model experiments and also analyzed observational data. Nagai et al. (1995) investigated the role of the Indian Ocean in the ENSO cycle. The results of both studies illustrate the key role of ENSO in generating interannual variability in all three tropical oceans.

Anomalies of sea surface temperature in the tropical Pacific force, via changes to the atmospheric circulation, anomalies of the same sign in the tropical Indian Ocean, as well as anomalies of opposite sign in the tropical Atlantic. The role of air–sea interactions in the tropical Indian and Atlantic Oceans is mainly that of an amplifier, by which ENSO-induced signals are enhanced in the ocean and atmosphere.

No evidence for zonal wave propagation around the globe, as proposed by Tourre and White (1996), was found in a modeling study by Latif and Barnett (1995). Consistent with observations, some eastward phase propagation was found in several model simulations. Anomalies of sea surface temperature in the Pacific induce a global response in winds. Because the different oceans exhibit different characteristics in response to low-frequency wind changes, the responses of the individual tropical oceans can by chance be timed to resemble a global wave, a coincidence which appears to be the case.

Middle Latitudes

While ENSO cycles have certain common and reproducible aspects in the tropics, their effects at higher latitudes are more variable. Large influences of ENSO, such as atmospheric wave trains propagating out of the tropics, can clearly be seen. However, the statistical reproducibility of these influences by the tropics on the middle latitudes is poor when the events are stratified solely according to whether or not a warm phase of ENSO is taking place. Modeling studies (admittedly with coarse atmospheric resolution, such as N.-C. Lau and Nath 1994) have indicated that variations of sea surface temperature in the tropics, characteristic of those found during the warm and cold phases of ENSO, affect the middle latitudes in recognizable patterns, and that midlatitude variations of sea surface temperature have only a small and indistinct effect. ENSO can have a variety of structures in the tropics, and each of these structures may induce a different reaction at higher latitudes. Furthermore, internally generated midlatitude variations may overwhelm the signal from tropical influences. These possibilities are briefly discussed here (see also, e.g., Trenberth 1995).

The primary source of ENSO is in the tropical Pacific. The observed global effects arise as the atmosphere transmits to higher latitudes the influence of anomalous heating in the tropics, driving changes to the large-scale overturning and thus anomalies in convergence and divergence. Regions of low-level convergence are mirrored in the upper troposphere by regions of divergence. The divergent outflow in the upper troposphere forces atmospheric Rossby waves (Sardeshmukh and Hoskins 1988), which propagate into the extratropics. The divergent outflow also provides an anticyclonic forcing when the Coriolis force acts on the outflow and, together with the advection of the Earth's vorticity by the anomalous divergent flow, helps create an anomalous Rossby-wave component (Sardeshmukh and Hoskins 1988, Rasmusson and Mo 1993). In the vicinity of the tropical heating, the response often takes the form of a pair of anomalous upper-tropospheric anticyclonic eddies that straddle the equator. The Rossby waves that emanate from these eddies can vary in wavelength and location according to the exact nature and scale of the forcing.

Several factors are important in determining the response of the extratropics to ENSO. First, rather small changes in both sea surface temperature and sea-surface-temperature gradients can greatly influence the locations of strongest convection and low-level convergence. The first-order response of the tropical atmosphere to anomalies of sea surface temperature is a shift in the location of organized convection. The second-order effect is a change to the character and intensity of the convection. Both effects lead to large anomalies in atmospheric heating, primarily through latent-heat release associated with precipitation. The spatial extent of the anomalous heating can vary considerably. In these respects, differences among the various ENSO cycles are large. These differences are apparent in the field of outgoing long-wave radiation in the tropics, where the response to sea-surface-temperature anomalies can be considered to be fairly direct. How much of this accounts for the huge differences among events in the extratropics is not yet known.

The second major factor in the response of the extratropics to ENSO comes from differences in the medium through which the forced Rossby waves propagate. These differences arise from the changing seasons and the changing location of the forcing. In addition, a random component arises from weather and weather-related variations, which dominate the extratropical circulation. Accordingly, the rather small influence on the extratropics from tropical forcing can be seen only if an average is taken over many synoptic events. The natural variability is therefore a form of unpredictable noise, while the signal caused by ENSO is regarded as potentially predictable and thus reproducible in a good model. For El Niño forcings, Kumar and Hoerling (1995) used an atmospheric general-circulation model to show that while the signal-to-noise ratio for the 200 mb geopotential height over the Pacific and North American extratropical region is roughly 0.2 for monthly means, it increases to about 0.6 for seasonal (3-month) means and to about 0.9 for 5-month means.

The above reasoning assumes a largely linear addition of the tropical influences on midlatitude patterns. This assumption is known from both observational studies (Horel and Wallace 1981) and modeling investigations (Hoskins and Karoly 1981) to be at least partially valid, although complications such as zonal inhomogeneities in the climatological background flow must be included (Branstator 1985). The Northern Hemisphere wintertime flow has a great deal of natural low-frequency variability arising from barotropic instability, and this same barotropic energy supply can be tapped by disturbances initiated by forcing localized to the tropics (Simmons et al. 1983). In addition, changes in the extratropical circulation immediately begin to change the jet stream and the associated storm tracks, so that the heat and vorticity fluxes caused by the transient eddies in the extratropics are also altered (Branstator 1995). The impact of these secondary changes, however, can be as large as or larger than the influence of the direct tropical forcing (Held et al. 1989, N.-C. Lau and Nath 1990, Kushnir and Lau 1992). Fortunately, some of these influences appear to be fairly systematic, although differing considerably by region, depending on the climatological background flow (Trenberth and Hurrell 1994). It is therefore possible to parameterize the effects of the changes in storm tracks (Branstator 1995).

Another view of the role of the tropical forcing on the extratropics was put forward by Palmer (1993) and Molteni et al. (1993). They note the very large natural variability of the extratropical circulation, but with the existence of certain preferred regimes where more persistent flow patterns recur. The persistent patterns are presumably associated in some way with the distribution of land and sea, and with the climatological planetary waves. The patterns indicate that one effect of forcing by tropical sea surface temperature is to alter the frequency of occurrence and stability of certain pre-existing regimes, but with only minor changes in the regimes themselves. Thus preferred "teleconnection patterns", such as the Pacific–North-American (PNA) pattern, may be excited. This helps explain why the response of the extratropical atmosphere to sea surface temperatures in the tropics can be highly nonlinear, as was found by Geisler et al. (1985).

The effects of ENSO are clearly global in extent. However, differences among ENSO cycles in the tropics are important and can become magnified in the extratropics. The anomaly patterns in the extratropics cannot be determined reliably by statistical means because the number of examples is not yet large enough to permit classification of the ENSO cycles into subtypes. Consequently, the best hope for predicting the remote influences of a particular phase of ENSO is good numerical modeling. Trenberth and Branstator (1992) discuss the criteria for performance of models if the simulations and/or predictions are to be useful. However, most models do not yet satisfy these demands. Improved models will allow the extratropical predictability that does exist to be better exploited.

ENSO and the Asian-Australian Monsoon

Although the heat source for the atmosphere associated with the western Pacific warm pool is of immense importance for, and has distinct and direct influences on, the global climate (see, e.g., Palmer and Mansfield 1984), it is not the only region of intense atmospheric heating. In addition to the latent heating over the warm pool and the Indonesian archipelago, there is a distinct migration of the heating maximum over south Asia during the spring and early summer. The heating variability is asymmetric with respect to the seasons; the deepest convection is located north of the equator in the western Pacific and over south and east Asia. During the austral summer, most of the convection remains fairly close to the equator. The annual migration of this heating maximum is associated with the Asian-Australian monsoon system. Thus, the annual cycle of the tropical climate system is driven by heating gradients of the coupled ocean–atmosphere–land system, and the monsoon system is an integral part of the Pacific Ocean climate. Furthermore, there are year-to-year variations in the intensity and location of the heat sources during the boreal summer. These variations relate directly to strong and weak monsoon seasons over Asia (Webster and Yang 1992).

A portion of the variability of the monsoons can be directly related to the influence of ENSO. Table 1 shows that of the 20 drought summer seasons over India (i.e., summers with less rainfall than one standard deviation below the mean), 11 of them were during ENSO warm phases (El Niño years); none was during a cold phase. On the other hand, no years during an ENSO warm phase were "very wet" years (i.e., years with summer more rainfall than one standard deviation above the mean), and 8 of the wet years were during ENSO cold phases. Thus, ENSO appears to explain a considerable amount of the rainfall variance over India (Shukla and Paolina 1983). However, Torrence and Webster (1996) point out that there were periods in which there was little ENSO variability and a correspondingly small covariance with Indian rainfall. From 1915 to 1960, variability in NINO3 (the anomaly of sea surface temperature averaged over the region between 5°N and 5°S, 150°W and 90°W) was quite

Table 2. Rain Conditions for India and Associated Phase of ENSO

	Total	During ENSO Warm Phase	During ENSO Cold Phase
Drought	20	11	0
Below-average rain	60	20	0
Above-average rain	61	2	16
Very wet	18	0	8

The relationship between the All-India Rainfall Index and the state of the Pacific Ocean for the period 1870 to 1991. See text for definitions of rain conditions. The table is an extension of one in Shukla and Paolina (1983).

small. In fact, the covariance between Indian rainfall and sea surface temperature in the western Indian Ocean was greater than the covariance between Indian rainfall and Pacific sea surface temperatures. The overall correlation between ENSO and Indian precipitation is dominated by two periods of very strong covariance, during 1890–1915 and during 1965–1980.

The waxing and waning of the relationship between the intensity of the monsoon and ENSO is tied to interdecadal variations of the planetary-scale climate system. Noting the failure of the relations between the SOI and Indian monsoon rainfall found earlier by Walker and his coworkers, Normand (1953) made the following assessment:

> *Unfortunately for India, the Southern Oscillation in June–August, at the height of the monsoon, has many significant correlations with later events and relatively few with earlier events.... The Indian monsoon therefore stands out as an active, not a passive feature in world weather, more efficient as a broadcasting tool than an event to be forecast.... On the whole, Walker's worldwide survey ended offering promise for the prediction of events in other regions rather than in India....*

Clearly, Normand was commenting from the perspective of nearly forty years of poor relationships between the monsoon and ENSO. Troup (1965) made a similar point in noting that there had been significant changes in the character of ENSO during the period 1915–1965. Balmaseda *et al.* (1995) and Torrence and Webster (1995) showed much later that the character of ENSO itself had also changed through the decades.

The monsoon and ENSO are parts of a climate system that evolves from decade to decade and that involves tightly coupled components. It is difficult to discern where processes begin. The monsoon system may couple to the extratropics through mechanisms involving Eurasian snowfall (see, e.g., Barnett 1984, 1985; Yang 1996). While examining the relationship between the monsoons and ENSO, Barnett (1984, 1985) and Barnett *et al.* (1989) detected a large-scale propagating surface-pressure signal that moved through the Indian Ocean region into the Pacific Ocean, with time scales greater than two years. The signal appeared to originate in the Asian region, and Barnett (1985) suggested that it was associated with Eurasian winter snowfall.

In an attempt to develop preliminary investigations into monsoon predictability and the relationships between monsoons and ENSO, the TOGA Program created the Monsoon Experimentation Group (MONEG). A principal mandate of MONEG was studying, using integrations of atmospheric general-circulation models, the effects of prescribed anomalies of sea surface temperature on the monsoon. To focus its study, MONEG chose to investigate the monsoon seasons of June–July–August during 1987 and 1988. As discussed by Krishnamurti et al. (1989a, b), these years were of particular interest because of their contrasting behavior. On one hand, 1987 was a severe drought year over both India and the African Sahel. On the other hand, 1988 was an above-average monsoon season for India, while rains over the Sahel were close to their long-term climatological mean. During these years, ENSO progressed abruptly from a warm phase in the summer of 1987 to a cold phase in 1988. This period was one in which the covariance between ENSO and the monsoon rainfall was strongest. The difference in sea surface temperature between the summers of 1987 and 1988 reached a maximum of 4 K in the eastern equatorial Pacific, flanked by differences of -1 K north and south of the equator. Other differences in sea surface temperature were evident around the globe, although the differences were much smaller in the Indian Ocean which tends to be in phase with the central Pacific.

In the first phase of coordinated experimentation, 17 atmospheric-modeling groups ran 90-day integrations with prescribed fields of sea surface temperature based on observations for 1987 and 1988. These are referred to as the MONEG integrations; full results from these integrations are discussed in WCRP 1992. The ability to simulate the basic monsoon climatology in a general-circulation model is not a trivial matter, and many of the models contributing to the MONEG-coordinated experiments suffered significant drift from the observed climate, often resulting in rather weak monsoon rainfall over some land areas (such as India). From the set of all MONEG integrations, it was apparent that a realistic simulation of the mean monsoon is a prerequisite to simulate interannual variability correctly, suggesting that the monsoon is an inherently nonlinear phenomenon.

Several models within the MONEG experimentation set (e.g., Palmer et al. 1992) were able to reproduce much of the observed coarse-grained interannual variability in the monsoon areas for 1987 and 1988, and the numerical experiments indicated that the skill was obtained from specifying the underlying (and evolving) anomalies of sea surface temperature, rather than from specifying the atmospheric initial conditions. This result is consistent with the Charney and Shukla (1981) hypothesis that most variability in the tropics is determined by variations in the state of the boundary.

The results (see, e.g., Palmer et al. 1992) showed overwhelmingly that variations in the monsoons were associated with observed variations in tropical Pacific sea surface temperatures. Observed sea-surface-temperature anomalies in the Indian and Atlantic Oceans did have some impact on the atmosphere over the Indian and Atlantic Oceans, respectively. Still, the role of interannual variability in the Indian Ocean on the Asian monsoon seems to be weaker than the remote effect of Pacific Ocean sea surface temperatures, at least during

ENSO years, and at least during the particular phase of the interdecadal variability that was studied. Since the completion of the MONEG integrations, Ju and Slingo (1995) have concluded that during cold phases of ENSO, warm anomalies of sea surface temperature in the tropical northwest Pacific may be of more direct importance in influencing the Asian summer monsoon than the cold equatorial anomalies of sea surface temperature in the central and eastern Pacific that determine the phase of ENSO.

In summary, it is clear that there is a predictable element in the summer monsoon of south Asia. At certain times during the last hundred years there are clear connections between ENSO and the strength of the Asian summer monsoon. Strong relationships exist between the Australian summer monsoon and ENSO with limited precipitation occurring during a warm phase and abundant precipitation during a cold phase. However, there are periods of low ENSO variance (e.g., 1920–1960) when there appears to be little connection between the monsoon and ENSO, and periods when the variability in the monsoon seems to lead ENSO variability. Questions remain as to the effects of the oceans on the monsoon during neutral phases of ENSO, or when ENSO has a different character from the 1987–88 episodes. Also, the effects of sea-surface-temperature anomalies on more regional quantities, such as country-wide seasonal mean rainfall, are largely unknown. From a practical forecasting point of view, these fine-grain issues are obviously of great importance.

Carbon Dioxide and ENSO

The eastern equatorial Pacific is a net source of carbon dioxide (CO_2) for the atmosphere. Carbon is exchanged between the atmosphere and the ocean in proportion to the difference of the partial pressures of carbon dioxide (pCO_2) in the atmosphere and in the ocean. In the eastern Pacific, the dominant control on pCO_2 is upwelling to the surface of carbon-rich waters from beneath the thermocline (Barber and Chavez 1983). The controls on pCO_2 in the western Pacific are by no means as clear (see Ishii and Inoue 1995), but seem to be dominated locally by temperature and salinity, both of which affect the solubility of CO_2, with very little effect from upwelling because the carbon-rich waters are too deep.

When ENSO moves into a warm phase, the thermocline in the east deepens and the connection of the cold, deeper, carbon-rich ocean to the surface is broken. The result is that the outgassing of CO_2 to the atmosphere ceases. The pCO_2 in the west seems to change with changes in salinity, but these variations do not lead to major changes in the outgassing of CO_2 to the atmosphere (Fushimi 1987). All other things being equal, the amount of CO_2 in the atmosphere should decrease during warm phases of ENSO. However, variations in the phase of ENSO drive changes in temperature and precipitation on land, and also affect the atmospheric concentration of CO_2. Although the correlation of variations in the SOI and atmospheric CO_2 (with variations in the SOI leading) was first reported more than two decades ago (Bacastow 1976), a quantitative understanding of the relative roles of the terrestrial versus the marine carbon pools in causing interannual fluctuations of CO_2 remains controversial.

The atmospheric anomalies of CO_2 observed during major ENSO warm events (1965, 1969, 1972, 1976, 1982, 1986, and 1991) are of order 1–2 ppmv; this should be compared to an annual-mean concentration now at 365 ppmv (increasing at a mean rate of approximately 1.5 ppmv/yr because of fossil-fuel combustion and deforestation). To put these values in perspective, note that a 1 ppmv increase in the global-mean atmospheric CO_2 level requires an input of 2 Gton (2×10^{15} g) of carbon (C; 1 Gton of C is equivalent to 3.7 Gton of CO_2). The balanced annual exchange of atmospheric CO_2 with the global ocean is approximately 100 Gton C/yr; a nearly comparable exchange of approximately 70 Gton C/yr occurs between the atmosphere and the terrestrial biosphere. Therefore, the largest CO_2 imbalances observed during climatically anomalous years of major ENSO phases represent only 1 percent fluctuations in the balanced exchange fluxes of atmospheric CO_2 with the oceanic and terrestrial biospheric reservoirs of carbon.

The 1982–83 ENSO warm event provided the first opportunity to quantify with direct measurements the major reduction of the sea-to-air CO_2 flux from the equatorial Pacific Ocean. Feeley et al. (1987) and Keeling and Revelle (1985) estimated the reduction to be approximately 0.6 Gton C. The reduction in flux results from the greatly reduced equatorial upwelling of deeper waters, which are strongly supersaturated with CO_2. The reduced ocean-to-atmosphere flux of CO_2 in the eastern equatorial Pacific during the TOGA years has been consistently confirmed by surface pCO_2 measurements throughout the development and decay of subsequent ENSO events in 1986–87 (Wong et al. 1993, Inoue and Sugimura 1992), and 1991–1994 (Feely et al. 1995, Inoue et al. 1996, Lefevre and Dandonneau 1992).

The pCO_2 of the surface in the western equatorial Pacific is not controlled by upwelling, because the thermocline is far from the surface and the carbon-rich waters cannot reach the surface by upwelling. There, the ocean-surface pCO_2 is almost the same as the atmospheric value, and very little outgassing of CO_2 occurs (Ishii and Inoue 1990). Furthermore, although there are measurable changes of pCO_2 with changes in salinity, there is very little temperature change in the west Pacific to drive variations in the surface pCO_2.

While a reduced ocean-to-atmosphere flux of CO_2 during ENSO warm events is now well established, this variation alone is not sufficient to explain the observed fluctuations in regional or global atmospheric CO_2 levels. The pattern and timing of the CO_2 variations during major ENSO warm events require a substantial, perhaps dominant, contribution from the terrestrial biosphere and the soil carbon pool. Ocean (only) general-circulation models seem unable to reproduce the magnitude or phasing of the observed atmospheric CO_2 fluctuations (Winguth et al. 1994).

The argument for terrestrial dominance is most forcefully made by co-interpreting the atmospheric time series of total CO_2 and the time series of isotopically substituted $^{13}CO_2$. (The $^{13}C/^{12}C$ isotopic ratio of carbon dioxide is substantially altered during photosynthetic exchanges of CO_2 with the terrestrial biosphere, but nearly unaltered by air–sea gas exchange.) Keeling has argued that the major pulse of CO_2 to the atmosphere at the conclusion of major ENSO warm events comes from the terrestrial biosphere, largely caused by the drought and fire in southeast Asia that accompany the failure of the monsoon, and that, more generally, the flux anomalies of the ocean and terrestrial biosphere during an ENSO warm event are large and of opposing sign (Keeling et al. 1989, Keeling et al. 1995). An independent record of ^{13}C by a different group shows a much smaller terrestrial contribution (Francey et al. 1995), and the topic remains controversial (Heimann 1995, Siegenthaler 1990).

The interpretation of the CO_2 anomalies accompanying the latest, and unusually long, period of warmth in the tropical Pacific (1991–1994, see p. 94) is complicated by the Mt. Pinatubo eruption (June 1991) and the subsequent cooling of the atmosphere because of the veil of stratospheric volcanic aerosol, which likely had major effects on the temperature-dependent oceanic and terrestrial carbon reservoirs.

THEORIES OF ENSO

Advances in the theory of interannual variability in the tropical Pacific related to the coupling between the atmosphere and ocean can be divided into four categories: the mechanism of ENSO (including attempts to verify some existing hypotheses), the irregularity of ENSO, the potential for interactions between the annual cycle and ENSO variability, and theories of ENSO prediction. Each of these is discussed below.

The Mechanism of ENSO

The fundamental idea underlying the current explanation of ENSO is that interactions between the atmosphere and ocean can give rise to instabilities of otherwise stable atmospheric and oceanic systems. This can be understood most simply by imagining a warm anomaly of sea surface temperature that creates surface winds, which in turn enhance the warmth of the anomaly. The anomalies of sea surface temperature and winds would grow in concert, and therefore be unstable. The problem is to identify those interactions and modes in the combined atmosphere and ocean system that have this behavior, and to see which (if any) correspond to ENSO.

The theoretical study of coupled modes began with a paper by Philander *et al.* (1984) that used simple, linear, "shallow water" models for both the atmosphere and the ocean. The scheme parameterized anomalies of atmospheric convection as proportional to anomalies of sea surface temperature, and parameterized anomalies of sea surface temperature as proportional to thermocline depth. Philander *et al.* found an unstable coupled atmosphere–ocean mode. The mode was eastward moving, with characteristics in the ocean very much like those of a free oceanic Kelvin mode, and with the atmospheric convection following the positive anomalies of sea surface temperature. Because, as we have seen, sea surface temperature anomalies by and large propagate *westward*, and eastward-propagating isotherms of total sea surface temperature travel with about a tenth the speed of a free Kelvin mode, this mode could not be identified with the ENSO signal. The Philander *et al.* paper was notable, however, in showing that coupled unstable modes can exist.

Hirst (1986) performed a more complete study of coupled atmosphere–ocean modes using simple shallow-water models. He found that long-period (several years) and slowly growing (several months) modes always exist, but that the nature of the coupled modes depends crucially on the mix of processes that control sea surface temperature. When anomalies of sea surface temperature are assumed to be simply proportional to anomalies of thermocline depth, then the unstable coupled mode resembles a free, eastward-propagating Kelvin mode in the ocean, with atmospheric convection following. When anomalies of sea surface temperature are changed by surface advection, the unstable coupled mode resembles a free, westwardly propagating Rossby mode in the ocean, with atmospheric convection following. However, when the rate of change of sea surface temperature was set proportional to thermocline depth (as a simple parameterization of the effects of mean upwelling), the nature of the coupled modes changed dramatically. A slow, eastwardly propagating mode arose, with ocean characteristics that did not resemble any free mode. While this coupled mode still could not be identified with ENSO, it provided a cautionary note to observationalists that propagating sea-surface-temperature anomalies in the ocean could not necessarily be identified with known free modes. One serious puzzle raised by Hirst (1988, Hirst and Lau 1990) was that these modes seemed oblivious to the existence of boundaries—they were identical in bounded and unbounded basins.

The coupled atmosphere–ocean model of Zebiak and Cane (1987) was built specifically to model interannual anomalies of sea surface temperature in the tropical Pacific. The key to the Zebiak and Cane model was the specification of monthly mean climatologies for both the atmosphere and the ocean, so that only anomalies about the specified climatology were calculated. The result of this coupled atmosphere–ocean model was a reasonable simulation of ENSO (see Figures 4–11 in Zebiak and Cane 1987). The model ocean consisted of a single baroclinic mode with a fixed-depth mixed layer. Anomalies of sea surface temperature were calculated with allowances for surface advection, mean and anomalous upwelling, and heat fluxes at the surface. The model atmosphere was a modification of the Gill (1980) model; it responded to anomalies of sea surface temperature and was also affected by the presence of *mean* convergences and divergences. In the coupled model, the anomalies of sea surface temperature grew in place and exhibited no particular direction of propagation. Clearly a mechanism for ENSO was contained in this model.

Battisti (1988) built a near replica of the Zebiak and Cane model to isolate the mechanism that produced ENSO-like oscillations. The model gave only regular ENSO-like oscillations but turned out to be appropriate for understanding a regular ENSO. (The differences between the Zebiak and Cane model and the Battisti model were explained only recently, by Mantua and Battisti (1995).) The oscillatory mechanism involved an intricate interplay among coupled instabilities, free equatorial modes, and changes in sea surface temperature. Consider a coupled instability in the eastern part of the Pacific such that, in the absence of boundaries, the anomaly of sea surface temperature would increase exponentially in time with a growth rate c and would be confined latitudinally to the region of the equator. The exponentially increasing surface winds, induced by the increasing sea surface temperature, would lie to the west of the sea-surface-temperature anomaly, as is usual in both models and observations. Because the winds cover a finite longitudinal extent, they tend to lower the thermocline to their east and raise the thermocline to their west, in accordance with the tendency of equatorial adjustment to bring the thermocline tilt in balance with the wind stresses (see Cane and Sarachik 1977). The downwelling of the thermocline to the east of the winds is consistent with the warming in the eastern Pacific. The upwelling (cooling) signal propagates freely westward with speeds and meridional structures characteristic of the lowest order symmetric Rossby mode, hits the western boundary, and returns behind wave fronts with speeds and meridional structure characteristic of Kelvin modes. After a time τ (the length of time it takes for the signal to return to the region of increasing sea surface temperature) the cooling signal, growing exponentially with an amplitude b as a function of its *retarded* time $t-\tau$, reaches the east Pacific and begins to erode the warming signal, eventually cooling the region. This is possible only when the remote signal has an amplitude that exceeds the initial signal.

The work of Battisti and Hirst (1989) formalized the above description. They found that it could be encapsulated in a single equation, the so-called delayed (or retarded) oscillator equation for the sea surface temperature at a particular place, which can be expressed as

$$\frac{dT(t)}{dt} = cT(t) - bT(t-\tau) - eT^3 ,$$

where the first two terms on the right-hand side form the linear part of the dynamics, $b>c$, and the final term allows the system to equilibrate at finite amplitude, with all coefficients real.

The coefficient c multiplies all terms of the thermal equation that locally change the temperature, including the damping by fluxes, while b multiplies only that term of the linearized thermal equation that arises from thermocline depth, i.e., mean upwelling of an anomalous vertical temperature gradient. Because the cooling signal travels a time τ before it shows up as sea-surface-temperature changes, it is effectively shielded from damping by surface fluxes and reaches the eastern Pacific undiminished. If the remote signal, multiplied by b, has a much greater ability to change sea surface temperature than the direct local effects, in the term multiplied by c, then growing oscillatory solutions can result. The cubic term is required for oscillatory finite-amplitude solutions. A somewhat similar argument is given by Cane *et al.* (1990).

The equation used by Schopf and Suarez (1988) was identical to the equation given above, with the crucial difference that they assumed $c>b$. For this condition, the linearized equation has only purely growing or decaying solutions, and any oscillatory solutions must arise from nonlinearity. Another way of looking at this is that while both of these models were formulated with the assumption that the coupled atmosphere–ocean mode was stationary and confined to the eastern basin, Schopf and Suarez postulated that the ocean dynamical adjustment time is large compared with the time scale associated with the air–sea interactions, and thus the coupled system is inherently nonlinear. The behavior of the retarded-oscillator equation as b changes magnitude from less than c to greater than c is discussed by Wakata and Sarachik (1994).

The relationship between the Hirst (1986, 1988) calculations of unstable modes, which did not at all resemble ENSO, and the ENSO-like mode in the Zebiak and Cane model remains unclear. The analysis by Hirst assumed a basic state that had no meridional structure for the upwelling. When the mean upwelling, with its rather narrow meridional distribution, was included, the mode found by Hirst was altered to become ENSO-like: it lost its eastward propagation and depended crucially on the boundaries (Wakata and Sarachik 1991b).

In a series of papers, Jin and Neelin (1993a, b) explored the instabilities that developed in a hierarchy of coupled models. Their study included an oceanic general-circulation model coupled to a statistical atmospheric model, and also a simplified model with an equatorial strip of the coupled atmosphere–ocean. Only the essential ocean thermodynamics from the Zebiak and Cane coupled model were retained. In these studies, Jin and Neelin explored the range of coupled atmosphere–ocean modes that exist in the parameter space covered by varying the coupling strength, the ratio of the ocean dynamical adjustment time to the time scale associated with the sea-surface-temperature changes, and the relative strength of the upwelling versus horizontal-advection terms in the ocean thermodynamic equation. By varying the parameters within realistic values, introducing increasingly complete ocean (sea surface temperature) thermodynamics, and tracking the eigenmodes, Jin and Neelin demonstrated how the extreme (idealized) eigenmodes that arise from the problem with a homogeneous basic state (see, e.g., Hirst 1988) give way to an increasingly realistic stationary dominant eigenmode, the delayed oscillator. It is clear from these papers, and the review by Neelin *et al.* (1994), that (unless some important physics is being neglected, such as feedbacks from clouds) the delayed-oscillator mode is rather robust and is likely to be the dominant, unstable, coupled atmosphere–ocean mode, at least in simple models.

The question remains: Is the retarded oscillator the actual mechanism for ENSO? In the simple models, it appears to be. However, in many important ways, the simple models do not faithfully represent nature. More complete (and complex) coupled atmosphere–ocean general-circulation models have been built, and the question is being asked of these models.

N.-C. Lau *et al.* (1992) reported on the tropical interannual variability simulated with a coarse-resolution coupled atmosphere–ocean general-circulation model, in which the Kelvin waves were significantly distorted because of the resolution (4° latitude) and numerical algorithms. Moreover, the latitudinal extent of the upwelling, known from theoretical constraints (Wakata and Sarachik 1991b) to be fundamental to the nature of the interannual variability, was not resolved. As a result of, and consistent with, all the results above, the interannual variability in this low-resolution model seems to be well described by a slow, westwardly propagating, destabilized Rossby mode, and not by a delayed-oscillator mode. Resolution sufficiently fine to resolve both the Kelvin mode and the meridional extent of the upwelling is clearly important for testing the delayed-oscillator hypothesis.

The delayed-oscillator theory for ENSO is remarkably consistent with the ENSO simulated by the Hamburg coupled atmosphere–ocean general-circulation model (Latif *et al.* 1993), and it is qualitatively consistent with the results from the Geophysical Fluid Dynamics Laboratory (GFDL) high-resolution coupled model discussed by Philander *et al.* (1992). The ENSO simulated with these models shares many similarities with the observed "canonical" ENSO (Rasmusson and Carpenter 1982). However, the interannual variability in a third coupled atmosphere–ocean general-circulation model, reported by Nagai *et al.* (1992), is rather weak compared to that observed, and does not seem consistent with the delayed-oscillator model of ENSO.

To what extent is the delayed-oscillator theory for ENSO supported by observations? Kessler (1990) has examined the observed wind-stress data, the anomalies of sea surface temperature, and the variability in the upper-ocean thermal structure obtained from the expendable bathythermographs launched by VOS. He concluded that the variability in the thermocline was consistent with that expected from the delayed-oscillator theory. Wakata and Sarachik (1991a) reached the same conclusion by examining the response of a "shallow-water" model to the long series of observed wind stress anomalies (Legler and O'Brien 1988).

While observations of the development and termination of the phases of ENSO are consistent with a delayed-oscillator theory of ENSO, there are clear inconsistencies between the observations and the delayed-oscillator theory during the onset of El Niño. Specifically, the strengthening of the trade winds that usually precedes the event is not a feature of the delayed-oscillator theory, nor of the simplified coupled atmosphere–ocean models from which the theory is derived. Li and Clarke (1994) used equatorial-wave theory and data from tide gauges to construct a record of the "observed" amplitude of equatorial Kelvin waves in the western Pacific. They correlated the reconstructed signal from the equatorial Kelvin waves with an index of the large-scale zonal-wind anomaly. Relying on those time series, they argued that the structure of the lagged-correlation curve was inconsistent with the delayed-oscillator theory of ENSO. The physics associated with the correlation structure reported by Li and Clarke was examined in a complementary analysis by Mantua and Battisti (1994), which used an ocean model forced with the observed anomalies of wind stress for the purpose of obtaining a signal of the equatorial Kelvin waves in the western Pacific (a signal that was highly correlated with the one derived by Li and Clarke using independent data and methods). Mantua and Battisti demonstrated that the delayed-oscillator theory did account for the termination of warm ENSO phases, but that the cold phases are not usually terminated by the ocean-adjustment process described by the delayed oscillator. The basic reason for this difference is that the ENSO phases are not nearly periodic. Nevertheless, the delayed-oscillator equation and theory can be viewed as the current paradigm for a regular ENSO cycle, remembering that such a regular cycle is not observed in nature. The delayed-oscillator mechanism for ENSO is internal to the coupled system of the atmosphere and the ocean. It is not possible to say that the cause of the oscillation resides in either the atmosphere or the ocean, nor is it possible to identify triggers in either: The cause lies in the interaction of the atmosphere and the ocean, and the trigger in some small perturbation.

Finally, it should be noted that the basic idea of coupled atmosphere–ocean instability has been questioned. Penland and Magorian (1993) and Penland and Sardeshmukh (1995) have hypothesized that the tropical Pacific atmosphere and ocean system is stable in a global sense, and that the ENSO variability is best thought of as a response of the tropical Pacific system to stochastic forcing. Thus, without external forcing of the system there would be no ENSO. This hypothesis for ENSO appears to be at odds with the all the aforementioned studies on unstable modes of variability. If the forcing hypothesis is correct, the intermediate-level coupled atmosphere–ocean models and the hybrid coupled models with a numerical ocean and a statistical atmosphere should demonstrate stable, not oscillatory, behavior.

Irregularity of ENSO

The irregularity of ENSO has long been known. It is reflected in the spectrum of the SOI which is peaked, but not spiked, near a period of 40 months (Rasmusson and Carpenter 1982). Investigators have hypothesized many reasons for the irregularity, or aperiodicity, of ENSO. These hypotheses can be loosely grouped into four categories: (1) "noise" internal to either the atmosphere or the ocean and independent of the coupling between the two media; (2) inherent nonlinearity in the coupled atmosphere–ocean system (or nonlinearity in the coupling itself); (3) changes in the external forcing; and (4) interactions between ENSO and the annual cycle.

One suggested cause for irregularity of ENSO is noise internal to the atmosphere or ocean. Specifically, investigators have discussed the impact of the atmospheric weather and other short-term disturbances on time scales from weekly to interannual. Explicitly or implicitly, weather is invoked for ENSO irregularity in the studies of Schopf and Suarez (1988), Zebiak (1989), Goswami and Shukla (1991), and Penland and Magorian (1993).

Nonlinearity in the climate system has also been suggested as a source of irregularity for ENSO. Munnich et al. (1991) examined a streamlined, nonlinear coupled model and found that irregular interannual variability can result from the coupling of the atmosphere and ocean. Recently, Mantua (1994) analyzed the Zebiak and Cane numerical model, demonstrating that the irregularity in its modeled ENSO was caused by an interaction between two unstable coupled atmosphere–ocean modes. He concluded that, although a single ENSO cycle in the model was well described by the delayed-oscillator theory, the irregularity is caused by deterministic, nonlinear processes in the model associated with the interactions between the unstable coupled modes.

Forcing from outside the tropics or long-time-scale variations of climate might also cause irregularities in ENSO. Various investigators (e.g., Barnett *et al*. 1989) have proposed that disturbances propagating into the tropical Pacific from the extratropics or the Indian Ocean can trigger or modify ENSO. Preliminary investigations by Meehl *et al*. (1993) and Graham *et al*. (1994) implicate the effect of changing concentrations of atmospheric carbon dioxide. Finally, the definitive study of Mass and Portman (1989) eliminates volcanic eruptions as a forcing of ENSO.

Xie (1995), however, has suggested that the irregularity in ENSO may be explained by a simple superposition of the annual "mode" and the ENSO mode, rather than an interaction between these two dominant modes. This hypothesis for the irregularity in ENSO is based on the very different physical processes that seem to govern the interannual ENSO mode and the annual cycle (see, e.g., Köberle and Philander 1994).

It is not clear, however, to what extent the spectrum of interannual variability in the tropical Pacific is stationary. Direct observational data is not adequate to define even the multidecadal variability in the tropical Pacific. The low-frequency modulation of ENSO could come about from nonlinearity in the coupled ocean–atmosphere–land system, involving either the upper ocean or changes in the (shallow) thermohaline circulation in the subtropics. Unfortunately, the present intermediate models of the coupled atmosphere–ocean system are inappropriate for such studies. Furthermore, it is not yet possible to run coupled general-circulation models for the thousands of model years that are required to examine these issues.

It may be possible to examine low-frequency variability of ENSO using data from paleoclimate studies. The available proxy data have been increasing in variety and geographic extent. For example, proxy data from tropical corals have been used to infer equatorial upwelling (Lea *et al*. 1989) and extrema in anomalies of wind stress (Shen *et al*. 1992a, b; Cole *et al*. 1993). Proxy indicators for the SOI have been derived from coral (Cole and Fairbanks 1990) and tree-ring data (Stahle and Cleaveland 1993). Finally, a connection has been suggested between ENSO and increased marine emissions of dimethyl sulfide at high southern latitudes (Legrand and Feniet-Saigne 1991).

Annual Cycle and ENSO

The annual cycle does not seem to be necessary for ENSO to be realized. Many models without annually varying insolation have proven to be successful at simulating interannual variations that resemble ENSO. These models include the intermediate coupled atmosphere–ocean models (see, e.g., Schopf and Suarez 1988, Mantua 1994) and coupled atmosphere–ocean general-circulation models (see, e.g., Philander *et al.* 1992, Latif *et al.* 1993). Furthermore, the annual cycle does not seem to be fundamental to the irregularity of ENSO in these models. One of the aforementioned models (Philander *et al.* 1992) was run without annually varying insolation, and the ENSO cycles that it produced occurred irregularly, nevertheless. However, ENSO is sufficiently linked to the annual cycle that it is possible to think of a canonical ENSO cycle, formed by compositing observations fixed to the calendar year (Rasmusson and Carpenter 1982).

Recently, simulations using coupled atmosphere–ocean general-circulation models that include annually varying solar forcing have been reported by Nagai *et al.* (1992) and Robertson *et al.* (1995a). In both of these studies, the annual cycles in the tropical Pacific of sea surface temperature, surface fluxes, and wind are qualitatively consistent with those observed. Both models develop coordinated interannual variations of the atmosphere and surface ocean in the tropical Pacific, albeit much weaker and only qualitatively similar in pattern to those observed. Overall, then, simulating interannual variability in the presence of a annually varying insolation continues to be a difficult problem. Some models have reproduced interannual variability of sea surface temperature and other models have reproduced the annual cycle, but simulation of the full spectrum of variability remains elusive. It should be recalled that the calculated annual cycle is an average over all the variability present in the system, hence it is not independent of interannual variability.

The processes most crucial for determining the annual cycle appear to be different from those most crucial for determining interannual variability. In particular, we now believe that interannual variability of sea surface temperature in the Pacific depends in an essential way on wind-driven variations of the thermocline, while heat fluxes at the surface act mainly to damp the interannual perturbations (Barnett *et al.* 1991). Annual variations of sea surface temperature depend critically on heat-flux variations at the surface, and thus in an essential way on radiative and cloud feedbacks (Köberle and Philander 1994). Low-level stratus clouds have presented particular problems. These clouds feed back positively on sea surface temperature, leading to more stratus accompanying low sea surface temperatures and less stratus accompanying high temperatures (when the stratus are replaced by trade cumulus). Existing models generally deal poorly with such clouds.

In some recent studies, investigators have postulated that annual-cycle forcing of the coupled atmosphere–ocean system is responsible for the irregularity of ENSO (Tziperman *et al.* 1994, Jin *et al.* 1994, Chang *et al.* 1994). These investigators studied interannual variability in numerical models by arbitrarily increasing either the annual cycle or the strength of the coupling between the atmosphere and ocean. Lacking annual-cycle forcing, the interannual variability in these models is periodic. As the coupling parameter is increased, the frequency of the interannual variability increases through a sequence of rational fractions of the annual cycle, but the ENSO-like cycle remains locked in phase with the annual cycle. The transition to higher frequencies is characterized by chaotic variability. In each of these studies, the physics associated with the ENSO mode appears to remain robust (e.g., in Jin *et al.* 1994 the ENSO mode is characterized by the delayed-oscillator physics) as the forcing is increased.

Several different coupled atmosphere–ocean models were used in these studies to argue for the importance of the annual cycle in producing irregularity of ENSO. However, even in chaotic regimes, under modest forcing, those models have spectra that are qualitatively dissimilar to the observed spectra. In addition, in none of these studies is the annual cycle actually forced by an accurate representation of the insolation cycle; rather, the amplitude of a parameter or process with an annual cycle oscillates. Nonetheless, these preliminary studies provide some insight as to how the annual cycle might interact with ENSO.

Optimal Growth of Disturbances and ENSO Predictability

The predictability of a deterministic system depends on the growth rates of the inevitably present initial errors. When these errors grow quickly, the system is less predictable than when initial errors grow slowly. The fastest-growing errors need not be normal modes of the system. In fact, for the general case of non-adjoint evolution operators, the fastest-growing errors change shape as they grow. Farrell (1990) has shown how the general problem of predictability can be identified with finding the optimals, i.e., the fastest-growing disturbances, of a given initial state. Blumenthal (1991) was the first to study theoretically the fastest-growing disturbances in the tropical Pacific. The finite-amplitude disturbances that grow most rapidly, in a system that is not linearly unstable, are said to experience "optimal growth". He examined output from a freely evolving simulation obtained from the coupled atmosphere–ocean model of Zebiak and Cane (1987). The optimal perturbations were determined by deriving a linear autoregressive (Markov) model from a reduced set of output quantities. In his study, Blumenthal found that when a fixed-amplitude perturbation is applied to the model output, the maximum growth over nine months is realized when the perturbation is applied in February. He also argued that the dependence on season of the disturbance growth is consistent with the seasonal dependence of forecast skill reported by Cane *et al.* (1986).

Penland and Sardeshmukh (1995) built a similar autoregressive model using the observed sea surface temperature. In contrast to Blumenthal, they reported that disturbance growth is not a function of season, and hypothesized instead that ENSO results from white-noise forcing of a globally linear system. Unfortunately, the results cannot be directly compared, for two reasons. First, the observed sea-surface-temperature data used to derive the Markov model are dominated by a record of unusually periodic ENSO events during the 1970s and 1980s. Hence, the regression model might have captured variability that is associated with ENSO, but does not in turn affect ENSO, and might lead to apparent (but false) predictability. In addition, because Penland and Sardeshmukh used sea-surface-temperature data but no thermocline data in their analysis, their hypothesis that the coupled system is globally stable required that the ocean dynamics equilibrate on time scales that are rapid compared with the changes in sea surface temperature. Analyses of observations by Clarke and Li (1994) and of ocean hindcasts by Mantua and Battisti (1994) indicated that this assumption is not valid for most of the ENSO cycle. Starting from a few months prior to the peak of a warm phase of ENSO, extending through that peak, and into the cold phase of ENSO, the observations are consistent with the physics of the delayed-oscillator model for ENSO. Thus, it is not surprising that Penland and Sardeshmukh found that the Markov model could not predict with skill the termination of ENSO phases.

The studies of Blumenthal (1991) and Penland and Sardeshmukh (1995) come to somewhat disparate conclusions concerning the predictability and stability of the coupled atmosphere–ocean system. Nonetheless, both studies suggest that the traditional approach to assessing the variability in the atmosphere–ocean system—identification of the fastest-growing normal modes—may not be the most instructive for understanding the predictability of the system. The linearized system need not be self-adjoint, either because of the coupling or because of the spatial variation of the background state, so the normal modes of the system need not be orthogonal. Thus, the fastest-growing normal mode cannot, in general, determine disturbance growth in the coupled models, and forecast skill is better explained by a study of the projection of the initial conditions on the optimal perturbations.

Working in a Larger Community

Some lessons of the TOGA Program are difficult to measure. These include changes in the culture of science and public awareness of science. The TOGA Program strengthened the bond between oceanic and atmospheric scientists, permanently altering some of the institutional and disciplinary arrangements in the oceanic and atmospheric sciences. It contributed to an awareness, in both the scientific community and the public at large, that there are aspects of seasonal-to-interannual climate variability that can be monitored, understood, and predicted—El Niño has become a household word.

The TOGA Program had an important influence on the culture of interdisciplinary research in the oceanic and atmospheric sciences. Knowledge of both the meteorology and oceanography of the tropical Pacific proved essential to unraveling the mysteries of ENSO. Although a small number of scientists have always been able to work across traditional disciplinary boundaries, TOGA taught a larger number how to do this. The TOGA Program increased the number of "amphibious" scientists, those who are equally at home with research in the atmosphere and in the ocean, or at least have a clear understanding of the nature of the coupled system. TOGA field programs, such as COARE, brought together oceanographers and meteorologists to formulate, plan, and implement new observational strategies for measuring and understanding processes that couple the ocean and atmosphere. Oceanic and atmospheric modelers collaborated on physically based coupled models for understanding and predicting ENSO. New strategies were developed for assimilating unprecedented amounts of ocean data to initialize coupled ocean–atmosphere prediction models. For the first time, ocean scientists were able to perform quantitative studies of the predictability of seasonal-to-interannual variations of large-scale ocean thermal and flow fields. The oceanographers and meteorologists who have collaborated within the TOGA Program for a decade intend to continue that collaboration in the fifteen-year post-TOGA Program, called GOALS (NRC 1994b).

Operational monitoring by the TOGA Observing System of interactions between the ocean and atmosphere encouraged cultural change. The TOGA Program promoted the free distribution of observational data and of numerical-model analyses of wind, surface and subsurface temperature, and other variables at sites remote from land. Many of the observations were made available within a day, and all of them within 30 days, of acquisition. The motivation for this requirement can be traced to scientists' frustration at their inability to observe the evolution of the 2- to 3-month onset phase of the 1982–83 El Niño. The introduction of satellite-communication technology greatly enhanced the ability to transmit and receive measurements from the VOS network and from moored and drifting buoys. Even though the 30-day oceanic operational time scale is much longer than the operational time scale for atmospheric observing systems, immediate distribution of TOGA data drove significant changes in the patterns of data collection and distribution in the ocean sciences. Basic oceanographic data had often been unavailable for periods as long as two years. Although meteorologists were accustomed to the rapid dissemination of operational atmospheric data, other information had often been viewed as proprietary, at least for a period of time. Research data, especially data for the oceans, and detailed results from numerical simulations usually were sequestered. The

TOGA Program, with its emphasis on analyses of the current state of climate systems, and on comparing the predictions of many modeling schemes with reality, encouraged and rewarded the immediate sharing of data. Many scientists learned that sharing their basic data, before it was fully analyzed, did not undermine their careers.

It is instructive to compare the TOGA Observing System in the Pacific Ocean combined with TOGA's experimental prediction models to the early attempts at establishing a numerical weather-prediction system for the United States. The implementation of the TOGA Observing System in the Pacific supported the development of numerous data-product bulletins and newsletters that began publication about a decade ago (see Appendix B for a listing). One of the first scientific uses of operational oceanographic information was demonstrated in August 1987 at the General Assembly of the International Union of Geodesy and Geophysics (IUGG), where an interpretative analysis of the June 1987 El Niño oceanographic conditions (sea surface temperature, near-surface currents, thermocline depth, sea-surface height, surface wind, and other variables) in the tropical Pacific was presented. Analyses and experimental short-range climate forecasts for the atmosphere and ocean are now routinely distributed to scientists and the general public through NOAA's Climate Diagnostics Bulletin.

In addition to changing the views and approaches of established scientists, the TOGA Program had an impact on the generation of climate researchers now in training. At present, at least ten major research universities have collaborative oceanic and atmospheric science programs. The TOGA Program recognized the need to entrain young researchers, and was instrumental in establishing the NOAA Postdoctoral Program on Climate and Global Change. In the past, postdoctoral fellowship programs had not generally been associated with research programs. (An exception is the Ocean Drilling Program, which also maintains a postdoctoral fellowship program.) The Climate and Global Change Postdoctoral Program awarded thirty fellowships by the end of 1994, covering a wide range of disciplines; ten of the fellowships were awarded in areas directly relevant to TOGA scientific objectives. A high percentage of the alumni of the program have received tenure-track positions at universities or senior research positions. It is envisioned that the NOAA Postdoctoral Program will continue to emphasize seasonal-to-interannual scientific objectives through its association with the GOALS Program.

Many of the scientific and institutional arrangements in existence prior to the TOGA Program were more of a hindrance than a help to such interdisciplinary activities. Oceanographers and atmospheric scientists found that their traditional working environments were inadequate for attacking large interdisciplinary problems such as those presented by ENSO. TOGA participants largely solved these problems. They developed a number of arrangements for coping with the realities of scientific and institutional activities that cross the boundaries of discipline and function. Achieving a balance among monitoring, modeling, empirical studies, and process studies required unprecedented cooperation among NOAA, NSF and other federal agencies when reviewing funding proposals and when filling crucial gaps during the implementation phase of the program. At the same time, meteorologists and oceanographers at universities and federal laboratories took a more active role in providing balanced scientific advice to the relevant federal agencies through the National Research Council's TOGA Panel and other structures. The coincidence of scientific interests and national priorities may allow the interdisciplinary collaborations developed during the TOGA Program to persist, and may help to define more clearly the relationship between climate research and societal needs. The development of the ability to predict seasonal-to-interannual climate variations has changed the ways in which many scientists think about their work and their obligations.

5. Organization

> The TOGA Program, both nationally and internationally, developed effective management, funding, and advisory structures. These structures served to plan and implement the TOGA Observing System, the making of short-term climate predictions, the TOGA COARE field program, and the research needed to support the TOGA Program. On the whole, scientists guided the program. Program managers largely followed the scientific advice.

The TOGA Program involved the acquisition and coordination, both nationally and internationally, of many different kinds of resources. These resources included continuous atmospheric and oceanic observations over the breadth of the Pacific Ocean, the regular operational atmospheric-observing system, new coupled atmosphere–ocean models, extensive supercomputer time, new data-assimilation techniques, new prediction methods, satellite data streams, and special data from process experiments. Although much less expensive than a large particle accelerator, TOGA was in the realm of "big science", at least for the earth sciences. As such, it required the establishment of effective management and advisory structures. Though TOGA was a large program, much of the scientific direction was developed "bottom up", from the work of individual investigators who supported the objectives of TOGA and obtained funding for projects related to the larger program.

The institutional arrangements for both the national and international TOGA Program were based on a generic tripartite management structure: a project office, an advisory mechanism, and a resource office. The project offices received advice from various panels, and received funding from the resource offices. The resource offices coordinated and gathered the funding. The project offices guided the performance of the program. Within the United States, the TOGA Project Office obtained advice from the NRC, relying on the efforts of the Advisory Panel for the TOGA Program (the TOGA Panel). The TOGA Project Office received resources through an interagency agreement under the U.S. Global Change Research Program (USGCRP). For the international effort, within the World Climate Research Programme (WCRP), the International TOGA Project Office received advice from the TOGA Scientific Steering Group and obtained resources through the Intergovernmental TOGA Board. The successes of the TOGA Program—especially in deploying and maintaining the TOGA Observing System, in fostering the expansion of short-term climate prediction, in recognizing the connection between observing and predicting, and in conducting the process studies needed to fill the gaps in our knowledge—were greatly aided by these national and international structures. These structures worked well, both individually and cooperatively.

U.S. ORGANIZATIONAL ARRANGEMENTS

The TOGA Program in the United States expanded the basic tripartite structure of advice, resources, and administration into the structure shown in Figure 14. Advice was provided primarily by the NRC, relying on its TOGA Panel. The resources were gathered in an interagency collaboration between NOAA, NSF, and NASA. The TOGA Project Office was housed within its lead agency, NOAA, and resided at NOAA's Office of Global Programs.

NRC's TOGA Panel

While the structure shown in Figure 14 is interrelated in intricate ways, it is perhaps simplest to start with the advisory mechanism. The advice and review provided to the TOGA Project Office was provided by the NRC. The NRC relied on the efforts of its Advisory Panel for the TOGA Program (TOGA Panel), which is under the Climate Research Committee (CRC). The task statement for the Panel was written in 1983. It ordered the following:

> Under the oversight of the Board on Atmospheric Sciences and Climate, through its Climate Research Committee, and the Board on Ocean Science and Policy, the Panel will:
> 1. Provide regular guidance on scientific policy and priorities to the U.S. TOGA Program Office and involved agencies on behalf of the Climate Re-

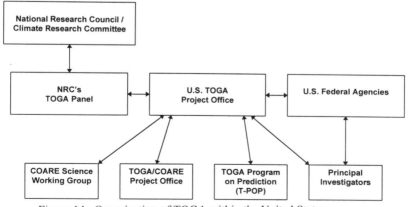

Figure 14. Organization of TOGA within the United States.

search committee (CRC). This should include Monsoon Climate Program activities.
2. Report regularly to the CRC on the Panel's involvement in U.S. TOGA plans and activities and to receive guidance from the CRC on TOGA matters in the context of the overall U.S. climate research program.
3. Advise on U.S. participation and provide the U.S. inputs to the international TOGA Program planning.
4. The members of the panel shall be selected to provide broad expertise in oceanographic and atmospheric research over all tropical oceans and the global atmosphere.

The TOGA Panel was established at the beginning of TOGA and has existed for the entire ten years of the program. It will disband upon completion of this report. Appendix A provides a list of the scientists who have given their time and energy to serve on the TOGA Panel. Questions have been raised about the number of panel members who were associated with the TOGA program, but it is not easy to determine who was part of TOGA. Although this report has largely taken an expansive view of the TOGA Program, an has included within the program domain almost all research efforts focused on ENSO or the prediction of seasonal-to-interannual climate variation, the formal designation of activities as part of the program was much more limited. If one takes this expansive view, then almost anyone who could provide for the NRC real expertise on seasonal-to-interannual climate variations was involved in the program. If one takes the narrow view that only activities supported by funds designated by federal agencies for TOGA were part of the U.S. TOGA Program, then most members of the panel did not have direct ties to the program.

The scope of the TOGA Panel's activities was limited to examinations and recommendations within the bounds of research related to ENSO. The panel did not suggest the balance of resources between those allocated for ENSO-related research and those allocated for other types of climate research. Examinations and recommendations on the place and magnitude of the TOGA Program within the broader climate-research efforts remained in the purview of the CRC. In fact, the science plan that helped launch the TOGA Program (NRC 1983) was prepared for the NRC by its more-broadly-constituted CRC, prior to the formation of the TOGA Panel. Recommendations for a follow-on program to TOGA, called the GOALS (Global Ocean–Atmosphere–Land System) Program (see chapter 7), were developed by the CRC (NRC 1994b), not the TOGA Panel.

The CRC (along with other parts of the NRC structure) has maintained oversight of the reports developed by the TOGA Panel, including this report.

Formal guidance and documentation of the TOGA Program were provided through published NRC reports. Following publication of the original scientific plan from the CRC, *El Niño and the Southern Oscillation* (NRC 1983), which suggested the establishment of a TOGA Program, the TOGA Panel produced a research strategy for U.S. participation in TOGA (NRC 1986). In 1989, the TOGA Panel initiated a workshop (Nova University 1989) to review the ocean observing system for TOGA, and then reported on the program's overall mid-life progress and directions (NRC 1990). The TOGA Panel later spun off a panel that reported on the Observing System and methods and problems for its maintenance (NRC 1994a). The TOGA Panel is now, in this report, summing up the accomplishments and legacies of TOGA. This series of reports have proven influential in attracting attention to the TOGA Program, in giving the scientific community confidence that the program was being well run and producing good science, and in ensuring that the community at large was being well represented by the TOGA Panel in its dealings with the U.S. TOGA Project Office and the participating federal agencies. The Panel also played a major role in helping the CRC define the science for TOGA's successor program, GOALS (NRC 1994b).

More frequent informal guidance was provided through presentations and discussions at meetings of the TOGA Panel. These discussions included the Director of the U.S. TOGA Project Office, program managers from participating agencies, and members of the research community participating in TOGA. Regular reports to the CRC (item 2 of the task statement) were facilitated by having the chair of the TOGA Panel serve *ex officio* on the Climate Research Committee. The Panel maintained contact with the international TOGA structure (item 3 of the task statement) by having its chair and other members regularly attend meetings of the international TOGA Scientific Steering Group.

Advice is valuable only when taken. Part of the success of the TOGA Program can be attributed to the willingness of the U.S. TOGA Project Office to take seriously the advice of the NRC, and the willingness of the TOGA Panel to take seriously the problems, opportunities, and funding limitations of the Project Office. NOAA program managers, more than most others, made an active effort to test their ideas in open discussion before the TOGA Panel to obtain an informal sense of the opinions of the research community on subjects for which consensus recommendations had not been prepared. The Panel did not intend to limit its interactions other agencies, but might have become more attuned to the concerns of NOAA. On balance, the good will and close relations between the Panel and the Project Office and other agency representatives has been one of the major reasons for the smooth running and success of the U.S. TOGA Program.

Although the job was not in its task statement, the TOGA Panel took responsibility for:
- providing a forum for federal agencies to work together, and also for aiding the process by which agencies contributed to TOGA and obtained benefits from it, according to individual agency needs;
- representing the larger TOGA community to the agencies and reporting back to that community any items of agency concern;
- presenting international problems to the agencies and to the TOGA Project Office and advising them on solutions;
- proposing the creation of new structures to carry out the TOGA Program more successfully; and
- documenting the progress of the TOGA Program in reviewed NRC reports, so that the larger community could see and benefit from the advances of the program, and also so that, through the NRC review process, the program could have a broader base of review.

The TOGA Panel also examined the merits of and then encouraged the execution of the large TOGA Coupled Ocean–Atmosphere Response Experiment (COARE) field program in the western tropical Pacific. Sometime after the midpoint of the program, the Panel recognized that the prediction aspects of the program were lagging and that no path existed for rapid progress (NRC 1990, p. 53). The Panel recommended the creation of the TOGA Program on Prediction (T-POP). The U.S. TOGA Project Office then commissioned a meeting that undertook to write a prospectus for a possible T-POP (Cane and Sarachik 1991). The prediction community used this prospectus as the basis for a sequence of proposals submitted to NOAA. Upon review and funding of the successful proposals, T-POP was formed. The TOGA Panel has maintained regular communication with T-POP participants, and has therefore been able to coordinate prediction research with requirements for the TOGA Observing System (see NRC 1994a) and with the process studies, especially COARE, needed to fill in the gaps in ENSO understanding and prediction.

Interagency Coordination of Resources

Resources were provided to the TOGA Program in the United States through an interagency agreement, which began informally. Interagency funding requests to the Office of Management and Budget and subsequent programmatic interagency implementation proved so successful that the process served as a model for the interagency coordination embodied by the USGCRP. TOGA also became part of the USGCRP. The U.S. government Subcommittee on Global

Change Research (now a subcommittee of the Committee on Environment and Natural Resources) coordinated interagency requests to Congress for TOGA funding. The several agencies involved then granted funding to individual principal investigators and to investigator groups.

The actual coordination of funding was achieved in a number of ways. In one, the program officers from the various agencies met informally from time to time and also attended the meetings of the TOGA Panel. In another, program managers attended panels of other agencies in which competitive funding was awarded. It was not unusual for proposals submitted to one agency to be funded by a different agency (or jointly funded) because of agreements reached as a result of cross-attendance at funding panels.

The major process study of the program, TOGA COARE, used a similar structure for management and funding. A project office was set up, funding was obtained through proposals submitted to the various agencies, and a science working group was established.

TOGA Project Office

The U.S. TOGA Project Office was housed at NOAA for the duration of the TOGA Program. Dr. J. Michael Hall was its first director. Dr. Kenneth Mooney was the second director ; he served until the end of the program. The Project Office coordinated the interagency support of individual principal investigators, organized science working groups (COARE and T-POP), kept the interagency budgets, and reported on a regular basis to the TOGA Panel. The director of the U.S. TOGA Project Office worked directly with the director of the International TOGA Project Office to ensure smooth integration and coordination of the national efforts with the international efforts. These coordinated efforts involved, for example, maintaining the TOGA Observing System, arranging international meetings and panels, and supporting numerical experimentation groups.

Science Working Groups

Two science working groups (SWGs) were organized by the TOGA Project Office. The TOGA COARE SWG was formed as an advisory mechanism for the massive COARE field program. The T-POP SWG was formed first as a provisional working group to define the initial outlines of the program (see Cane and Sarachik 1991). It later served as a gathering of principal investigators to coordinate the various prediction activities involved with TOGA.

COARE was such a large, comprehensive, and diverse experiment that it required its own management structure. Its advisory mechanism within the

United States was the COARE SWG, its resource office was an interagency collaboration, and it was implemented by a TOGA COARE International Project Office. The U.S. COARE SWG, assembled by the U.S. TOGA Project Office in 1988, was comprised of ten scientists, meteorologists and oceanographers from both universities and laboratories. Roger Lukas and Peter Webster co-chaired the SWG. An informal working subgroup structure was developed, with groups on large-scale circulation and waves in the oceans, large-scale circulation and waves in the atmosphere, atmospheric convection, ocean mixing, air–sea fluxes, and modeling. The two groups on large-scale circulation and waves provided the primary linkage to the overall TOGA scientific perspective. This structure proved crucial to the successful development of COARE.

The functions of the U.S. TOGA COARE SWG were to refine the scientific objectives, develop a research strategy, design the experimental approach, identify resource requirements (observing platforms, facilities, and other support), examine mechanisms for implementing the experiment, make recommendations about the coordination of research proposals, and establish, in cooperation with the TOGA Scientific Steering Group (SSG), the international framework for the experiment. The SWG worked closely with international colleagues to write the COARE Science Plan (WCRP 1990) and to develop the Experiment Design for COARE (TCIPO 1991), and worked with the TOGA COARE International Project Office (TCIPO), after it was established in 1990, to develop the Operations Plan for COARE (TCIPO 1992).

The SWG was not intended to last through the field phase and into the analysis and synthesis phases of TOGA COARE. This was, perhaps, an oversight. It became clear that continuing guidance and oversight were needed to achieve climate-prediction benefits from the substantial investment in COARE. With the formal end of TOGA, the full analysis of the data from TOGA COARE is in jeopardy. The 1994 COARE data workshop in Toulouse, after considerable discussion, determined that to produce a final, quality-controlled data set, funding will be required through at least 1998 (TCIPO 1995). This schedule is consistent with experience from previous large field programs. However, it is not obvious how such funding can be sustained, especially because resources devoted to TOGA will not necessarily be continued intact to its successor program, GOALS.

Coordination with International TOGA

No program dealing with the climate of major parts of the globe could possibly be accomplished without international coordination and cooperation. The U.S. TOGA Program had a significant involvement with international organizations. In order to assure smooth coordination between the national and international organizations, the U.S. TOGA Project Office seconded several people to work in the International TOGA Project Office, and also partially funded the staffing of the international office.

The NRC's CRC became the official liaison between the U.S. science community and the WCRP because one of the sponsors of the WCRP is the International Council of Scientific Unions, to which the National Academy of Sciences[*] belongs. The TOGA Panel was designated by the CRC to serve as its liaison with the International TOGA SSG.

In addition to the multinational arrangements, the United States entered into bilateral agreements with other countries to develop and maintain certain observing-system elements. In particular, the United States sustained a bilateral agreement with New Zealand for the maintenance of the upper-air station at Canton Island (now closed). Agreements were made with France and Australia for the maintenance of lines of ship-launched expendable bathythermographs. The United States entered into agreements with the People's Republic of China and with Indonesia for collaboration in the western Pacific and in measurements for TOGA COARE.

INTERNATIONAL ORGANIZATIONAL ARRANGEMENTS

The International TOGA Program was coordinated under the auspices of the WCRP. Several specially created entities then developed the program specifics for TOGA. Resources were arranged by the Intergovernmental TOGA Board. The TOGA SSG served as the advisory body and the International TOGA Project Office organized the implementation (see Figure 15).

TOGA as an Integral Part of the WCRP

Following directly from the highly successful Global Atmospheric Research Program (GARP), the WCRP was initiated at the end of 1979. The initiation was formalized by an agreement between the World Meteorological Organization (WMO, a specialized agency of the United Nations), and the nongovernmental International Council of Scientific Unions (ICSU). The Joint Organizing Committee (JOC) for GARP subsequently metamorphosed into the Joint Scientific Committee (JSC) for the WCRP. The major objective of the WCRP is to determine the extent to which climate can be predicted and the extent of man's influence on climate. WCRP initiatives are concerned primarily with time scales from several weeks to several decades. An examination of the WCRP, with an emphasis on U.S. participation and including a chapter on the

---

[*] The NRC serves as the operating arm of the National Academy of Sciences.

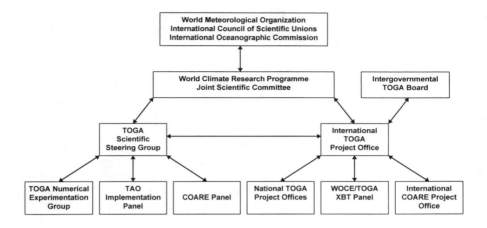

Figure 15. International organization for TOGA.

role of TOGA within the larger international program, was the subject of a 1992 NRC report.

At the inception of the WCRP, the ICSU Scientific Committee on Oceanic Research (SCOR) and the UNESCO (United Nations Educational, Scientific, and Cultural Organization) Intergovernmental Oceanographic Commission (IOC) established the Committee on Climatic Changes and the Ocean (CCCO) to facilitate the involvement of the international oceanographic community in WCRP planning. The CCCO worked in collaboration with the JSC until the CCCO was disbanded at the end of 1992. In 1993, the IOC joined WMO and ICSU as a joint sponsor of the WCRP, and the membership of the JSC was expanded accordingly.

Early consultations between the JSC and the CCCO led to the formation in 1981 of a joint JSC/CCCO study group on the interannual variability of the tropical oceans and global atmosphere. The study group sought to build on national efforts, most notably the TOGA Program that was then evolving in the United States, to define a program on interannual variability. The joint JSC/CCCO Study Conference on Large-Scale Oceanographic Experiments in the WCRP, held in Tokyo during May 1983, afforded the first opportunity for extensive international discussions of TOGA as a possible element of the WCRP. The conference concluded that the TOGA Program was feasible, that it should be an integral part of the WCRP, and that it should be organized as soon as possible.

The conference also recommended that the World Ocean Circulation Experiment (WOCE), then developing in the United States, should be expanded as a major activity within the WCRP. Together, TOGA and WOCE have constituted the major oceanographic efforts of the WCRP. For the most part, they complemented each other. Overlapping requirements in their respective implementation plans for a number of oceanographic observations encouraged some coordinated efforts, notably for certain expendable-bathythermograph, drifting-buoy, and tide-gauge data, together with the associated data-management activities. For this reason, close liaison was maintained between the two international programs at the levels of the scientific steering groups and project offices. This cooperation extended to joint sponsorship of the TOGA/WOCE XBT/XCTD (expendable bathythermograph / expendable conductivity-temperature-depth probe) Program Planning Committee and the WOCE/TOGA Surface Velocity Program Planning Committee.

Scientific Steering Group (SSG)

In early 1983, a joint JSC/CCCO TOGA SSG, chaired by Adrian Gill, was established. The TOGA SSG formulated the overall international program. It developed the concept of the TOGA Observing System and provided scientific guidance for its implementation. The SSG assumed responsibility for leading and coordinating the work of the three CCCO Pacific, Atlantic, and Indian Ocean Climate Studies Panels. As TOGA developed, the SSG ensured the exchange and analysis of TOGA data and the dissemination of scientific results. To provide overall scientific guidance for COARE, the TOGA SSG established the international TOGA COARE Panel, which worked closely with the U.S. COARE Science Working Group.

The SSG met for the first time during August 1983 at the Max Planck Institut für Meteorologie in Hamburg to initiate plans for the preparation of the TOGA Scientific Plan. It organized the International Conference on the TOGA Scientific Program, which was held in Paris during September 1984. The first draft of the Scientific Plan for TOGA was distributed at the Paris conference. The final version was published by the WCRP in September 1985. Throughout its lifetime, the SSG maintained a close relationship with the U.S. TOGA Panel.

Scientific advice at the international level was also provided by the International TAO Implementation Panel. It was convened to coordinate the Tropical Atmosphere/Ocean (TAO) Array, a collection of moorings in the tropical Pacific that served as a cornerstone of the TOGA Observing System. The TAO Implementation Panel brought the various users of TAO array information together. Progress and problems with the array and its instruments were assessed and reviewed, with an eye towards improving the analysis of the TAO observations and commenting on the adequacy and accuracy of the array itself.

International Project Office

At its first meeting, the TOGA SSG formally established the International TOGA Project Office (ITPO) and accepted the U.S. offer to host the ITPO. The ITPO prepared detailed plans for the various observational components of TOGA, for data processing, and for the production of reference data sets. It also assisted with the preparation of TOGA documents and with other organizational tasks.

The ITPO was located in Boulder, Colorado, from 1984 until its move to Geneva in 1987. When in Boulder, the ITPO was supported primarily by the United States, with additional personnel seconded by France and India. Rex Fleming and James Lyons served as the first two directors, while the office was in Boulder. When in Geneva, the ITPO was supported by the United States, Canada, the United Kingdom, and the WMO, with personnel seconded by France. John Marsh served as director while the office was in Geneva. The support of the ITPO by the United States was a U.S. contribution to the management of the TOGA program.

The ITPO (Boulder) produced the first edition of the International TOGA Implementation Plan in 1985. The second edition followed in 1987. Further editions, which reflected significant contributions of resources from many nations, were produced by the ITPO (Geneva) in 1990 and 1992. The Implementation Plan relied on such observing systems as the WMO's World Weather Watch (WWW) and the IOC's Global Sea-Level Observing System (GLOSS). In addition to these ongoing efforts, and the special efforts of the United States, contributions came from Australia, France and Japan for the VOS XBT program; contributions came from France, Japan, Korea, and Taiwan for the TOGA TAO array of moorings in the western Pacific; contributions came from Australia, France, Japan, Korea, and Taiwan for the Surface Velocity Program; and Ecuador, the Maldives, Papua New Guinea, and the Republic of the Yemen operated key TOGA upper-air stations.

In accordance with the TOGA Data Management Plan, which was an integral part of the International TOGA Implementation Plan, TOGA data centers were operated by European Centre for Medium-Range Weather Forecasting, France, India, and the United Kingdom. These data centers worked in concert with the TOGA Global Sea Surface Temperature Data Center and TOGA Sea-Level Data Center established in the United States (at the Climate Analysis Center of the National Meteorological Center* and the University of Hawaii, respectively).

---

* Both of these have since been renamed, to the Climate Prediction Center and the National Centers for Environmental Prediction, respectively.

A separate project office was established in Boulder for the large COARE field program. The office moved to Townsville, Australia, during the Intensive Observation Period. The TCIPO supported the ongoing planning for TOGA COARE and coordinated international implementation. The TCIPO worked with the TOGA COARE Panel for the purpose of preparing and then updating the international TOGA COARE Operations Plan (TCIPO 1992), to coordinate national commitments of resources in relation to the implementation plan, to develop and implement a logistics support plan for the field phase of COARE (TCIPO 1993), to develop and implement a specific data management plan for COARE, and to coordinate information among scientists, national agencies, and appropriate international bodies. The TCIPO was headed by David Carlson, who saw to the efficient operation of the office, provided general leadership to conduct the COARE field phase, and oversaw the data-management plan. The TCIPO published a summary of the observations made during the Intensive Observing Period of COARE. The office closed in June 1996 after the last director, Richard Chinman, finished supervising final data processing and distribution.

Intergovernmental TOGA Board

Following recommendations from the first informal planning meeting on the WCRP, held in Geneva during May 1986, the WMO and IOC established the Intergovernmental TOGA Board in 1987. The principal objectives of the board were to review the requirements for the implementation of TOGA, to provide a forum for intergovernmental consultations to ensure coordination of national resources that might be applied to the TOGA Program, to review progress made in implementing TOGA, to identify gaps that might appear in the implementation of TOGA, and to take action, as appropriate, to fill these gaps. Fourteen nations closely allied with TOGA were originally represented on the board; they were joined later by four others. The final national representation consisted of Australia, Brazil, Canada, Chile, China, Ecuador, France, Germany, India, Indonesia, Japan, Mauritius, New Zealand, Pakistan, Peru, the United Kingdom, the United States, and the U.S.S.R.

At the second meeting of the Intergovernmental TOGA Board in December 1988, the U.S. representative informed the board of the U.S. planning undertaken COARE. The board formally adopted COARE as an integral part of TOGA. The TOGA COARE Science Plan was published by the WCRP as an addendum to the Scientific Plan for TOGA (WCRP 1990). In addition to the United States, many nations (notably Australia, China, France, Japan, Korea, New Zealand, Taiwan, and the United Kingdom) made significant contributions of personnel and equipment to COARE. These resources enabled a far more ambitious field program to be designed than would have been possible had U.S. assets alone been involved.

At its final meeting in April 1993, following discussions initiated by U.S. representatives at previous meetings, the Intergovernmental TOGA Board noted the strong statements of support for and interest in directly participating in various aspects of an International Research Institute for Climate Prediction (IRICP). This proposal had emanated from a NOAA study group with international representation. The TOGA Board requested that the United States continue to take the lead in developing an IRICP proposal by convening a broad high-level meeting of interested nations to consider the creation of such an institute and to begin its implementation. These efforts are described in Chapter 7.

6. Applications of ENSO Prediction

As the skill for predicting sea surface temperature in the Pacific improved during the latter half of TOGA, attention was directed towards applying these predictions for economic and societal benefit. Physical and social scientists opened a dialogue during TOGA on how to develop and apply forecasts of short-term climate variations. Several of the tropical nations directly affected by ENSO variations have begun applying the fledgling ENSO predictions.

The scientific objectives established at the outset of the TOGA program focused solely on observing and modeling the coupled atmosphere–ocean system. Although it was anticipated in 1985 that observations and coupled-modeling efforts under TOGA would lay the foundation for operational prediction of short-term climate variations, the time when this might be practical could not be foretold. Assessment studies of the societal impacts of ENSO were not an integral part of TOGA at the beginning. By the mid-point of the program, however, scientists had demonstrated significant skill in predicting aspects of ENSO—sufficient skill for the forecasts to have beneficial value for society.

The demonstration of significant skill led the TOGA Panel to recommend the establishment of an experimental center for short-term climate prediction (NRC 1990). As the prospects for skillful and regular ENSO forecasts became more of a reality, a demand was generated for quantitative assessments of the socioeconomic impacts of ENSO. Similarly, it became necessary to understand how to apply ENSO forecasts and to formulate strategies for mitigating the more harmful aspects of climate variability. During the second half of TOGA, applications were emphasized as a necessary adjunct to the modeling, observational, and process-study components of the program. It was not a coincidence that this increased emphasis on ENSO applications evolved in tandem with a related emphasis within the USGCRP (CEES 1993) on the human dimensions of global climate change.

Prior to the start of TOGA, the most well-studied societal impact of El Niño was the collapse of the Peruvian anchoveta fishery in 1972–1973 (Glantz and Thompson 1981). The effects of El Niño along the west coast of South America received widespread attention. What was once thought to be a regional climate problem was seen to have global economic implications in the form of rising prices for fish meal. Examinations of lead and lag relations between the Southern Oscillation Index and regional climate anomalies, particularly droughts, had already established global connections. However, decision makers in Peru were not aware of this climate information that might be useful for their planning of agriculture and aquaculture. The global socioeconomic effects of ENSO did not attract intense interest until after the severe conditions of 1982–1983. Just as the 1982–83 climate anomalies mobilized the meteorological and oceanographic scientific communities under TOGA, so did the related human suffering draw the attention of the social, economic, and political communities. Drought in Australia, Indonesia, northeast Brazil, and southeast Africa, along with flooding in Ecuador, Peru, southeastern South America, and the western and southern tier of the United States, had adverse effects on social conditions and some economic sectors. Economists, social scientists, anthropologists, and politicians came together with the physical scientists to study the global societal effects of ENSO.

One of the legacies of the TOGA program is the dialogue that was established between the physical and social sciences (Moura 1992). This exchange of ENSO information was critical to the ultimate success of the program. Prior to TOGA, most information on ENSO impacts was anecdotal and derived from the popular press. TOGA helped to sponsor and support several forums for those with a common interest in ENSO. Sociologists, economists, and politicians were informed of the advances in and limitations of experimental ENSO forecasts. Climate scientists learned of the needs and problems of policy and decision makers. Education and training programs were geared to the application of ENSO forecasts for participants from tropical and developing nations. Collaborative arrangements were initiated between potential providers and users of ENSO forecasts. The governments of countries that are most directly affected by ENSO (e.g., Peru, Brazil, and Australia) now routinely take experimental ENSO forecasts into account when making decisions.

It was within this context of being able to provide societally relevant climate information to the countries affected by ENSO that a proposal (Moura 1992) for an International Research Institute for Climate Prediction (IRICP) arose. In June 1992, at the United Nations Conference on Environment and Development (UNCED), the Bush administration committed the United States to implementing "a pilot project to demonstrate the operating concepts embodied in the plan and invited government officials and scientists from all interested nations to join in developing an International Research Institute for Climate Prediction" (OMB 1992).

DEVELOPMENT OF APPLICATIONS AND ASSESSMENTS

ENSO forecasts are societally relevant because the agrarian sectors in tropical countries depend critically on the water received during the rainy season. In the tropics, precipitation and sea surface temperature are tightly coupled, so useful precipitation information can be derived from forecasts of sea surface temperature. Information on the probability of decreased or enhanced rainfall on seasonal-to-interannual time scales offers some countries the opportunity to take intelligent action.

The impacts of ENSO are varied. They depend on whether a region is rural or urban, whether the economic sector is agricultural or industrial, and whether the land holdings are small or large. The form of response to ENSO will vary according to the past history of the forecasts, the ability or desire of the government to intervene, and the acceptance of the information by the citizenry. Some societies are averse to risk and will implement mitigation strategies, while others may choose not to do so. For example, if upcoming precipitation patterns or tendencies are predicted with sufficient accuracy and a drought is forecast, a farmer can substitute drought-resistant crops, usually reducing yield but avoiding disaster. Similarly, a commodities broker has the option of obtaining supplies from another portion of the globe if a particular crop is in danger.

During the course of the TOGA decade, physical scientists, social scientists, economists, and decision makers from industry and government came together to analyze a range of factors influencing responses to ENSO effects by country and region. This entailed assessing the consequences of ENSO, discussing the application of ENSO forecasts, and judging the economic value of ENSO predictions. These assessments included studies of agricultural production, coastal fisheries, emergency preparedness, systems for early warning of drought and famine, strategies for hydroelectric power generation, insurance concerns, and mitigation of forest fires. A series of workshops proved very successful in bringing these seemingly disparate communities together to work on topics of mutual concern. The series* included:

Societal Impacts Associated with the 1982–83 Worldwide Climate Anomalies, November 1985, Lugano, Switzerland;

Workshop on the Economic Impact of ENSO Forecasts on the American, Australian, and Asian Continents, August 1992, Tallahassee, Florida;

Workshop on ENSO and Seasonal-to-Interannual Climate Variability: Socio-Economic Impacts, Forecasting, and Applications to the Decision Making Process, September 1992, Fortaleza and Florianopolis, Brazil;

Pacific El Niño–Southern Oscillation Applications Workshop, October 1992, Honolulu, Hawaii;

---

* Informally published proceedings of most of these workshops were prepared and can be obtained from the NOAA Office of Global Programs.

Usable Science: Food Security, Early Warning and El Niño, October 1993, Budapest, Hungary;

Inter-American Institute Workshop on ENSO and Interannual Climate Variability, July 1994, Lima, Peru;

Regional Workshop on Research and Application of Climate Forecasts in the Decision Making Process in Northern and Northeastern South America, September 1994, Fortaleza, Brazil;

Regional Workshop on Research and Application of Climate Forecasts in the Decision Making Process in Southeastern South America, September 1994, Montevideo, Uruguay; and

Workshop on Short-Term Climate Forecasts and their Applications for Social and Economic Benefit and Sustainable Development, November 1994, Bali, Indonesia.

The cross-disciplinary communication fostered by these meetings heightened awareness of the relationships between climate fluctuations and society. It became apparent that scientists in developing nations were on the verge of being able to advise their governments on the value of short-term climate predictions. Numerous examples and case studies made it clear that a well developed strategy would be needed to produce accurate and socioeconomically relevant climate information. These workshops, held during the last half of TOGA, helped to identify the needs of the economic sectors most affected by short-term climate variations, areas in need of further work if ENSO forecasts are to be useful for mitigating economic and social impacts, and methodologies for the systematic production and dissemination of climate-forecast information. Emerging from these exchanges was an appreciation of the issues encountered in applying ENSO forecasts, as well as several examples where governments were actually taking action on the basis of experimental short-term climate predictions.

During the workshops, it was recognized that economic assessment exercises were being done on a case-by-case basis. A systematic quantitative evaluation of the global economic impact of ENSO was called for. It was noted that programs are needed to improve the confidence in ENSO forecasts by means of education, data exchange, improved regional-model accuracy, a focus on probability forecasts, and support for near-real-time, sustained environmental monitoring. The fact that the societal response mechanisms differ from country to country means that communication and education are key to the successful application of ENSO forecasts. There is a fundamental obligation to consult with, and understand the needs of, the users and consumers of the information. The education that is needed to accomplish this goes in both directions. In order to be useful, ENSO forecasts, bulletins, and advisories must be tailored for users on the basis of input from them. The end users must also be apprised of the limitations of the forecasts. Regular production of forecasts is needed to build confidence and establish rapport between forecasters and users. Decision makers must be informed of the value of today's climate information and how it might be used in the future.

In response to this need for education and training, and to the U.S. proposal to UNCED for an International Research Institute for Climate Prediction pilot project, an applications and training pilot project was established at the Lamont-Doherty Earth Observatory in 1993. Approximately 29 participants from 17 countries were trained as of the end of 1994. The trainees learned about current capabilities in climate modeling, and helped to develop practical applications for their home country that link model forecasts with regional impacts.

APPLICATIONS OF REGIONAL ENSO FORECASTS

At the end of TOGA, ENSO forecasts and their applications were performed by small research groups in developed countries on an *ad hoc* basis. These groups lacked the infrastructure required to incorporate all relevant information into forecasting schemes. Research groups did not have the resources to adapt and disseminate the forecasts on the regional scale. Notwithstanding these impediments, many countries now recognize the value of short-term climate forecasts. Countries such as Peru and Brazil provide the best examples of how this forecast information has been used by policy makers to mitigate the socioeconomic effects of ENSO. The successes achieved in South America and elsewhere have demonstrated that a strong partnership is needed between developed and developing countries if all are to benefit from the new knowledge of ENSO predictability.

Peru

In Peru, the government plays an important role in setting agricultural policy. Every year an agricultural plan must, by law, be developed. This plan encompasses irrigation practices, prices of seed and fertilizer, agrarian loans, crop subsidies, and technical assistance. Following the 1982–1983 El Niño, the gross national product and the gross value for the agricultural sector dropped drastically. However, because of the application of predictive information, agricultural production did not decrease following the 1986–1987 El Niño (Lagos and Buizer 1992). The growing season of 1986–1987 was the first time the agricultural plan had been influenced by the prediction of an upcoming warm event. In late 1986, government officials and nongovernmental agrarian organizations were advised by Peruvian scientists that the observation and modeling efforts of

TOGA investigators were forecasting the coming of El Niño. The prediction of a warmer and wetter growing season helped determine the appropriate combination of crops. Two of the primary crops of the northern coastal region of Peru, rice and cotton, are very sensitive to the amount and timing of rainfall. Rice requires a large volume of water and relatively warm conditions during its growing phase. Cotton has deeper roots and is capable of greater yields during years of light precipitation. By using ENSO-based forecasts as a basis for crop selection, Peruvians have been able to sustain the gross value of their agricultural sector, increasing it 3 percent in 1987 in spite of the moderate 1986–1987 El Niño. In contrast, it decreased by 14 percent in 1983 following the 1982–1983 episode before forecasts were available.

Northeast Brazil

Northeastern Brazil, or the Nordeste, offers another example of the advantageous use of interannual climate predictions for agricultural planning. This semi-arid region, home to 30 percent of Brazil's population, encompasses 18 percent of the country's area (Magalhaes and Magee 1994). The Nordeste has a long history, going back to the colonial era, of social upheavals and migrations tied to droughts. In this region of nearly 45 million inhabitants, one in five Nordestinos emigrates to another portion of the country, often the Amazon, to escape drought. During modern times, drought in the Nordeste has required the federal Brazilian government to devote considerable resources for ameliorating the drought conditions in order to forestall further emigration.

In the Nordeste state of Ceara, the government has demonstrated that population displacements can be reduced by assisting subsistence farmers. The government in this state has become quite sophisticated and active in using forecasts related to ENSO. The state-supported Ceara Foundation for Meteorology and Hydrological Resources (FUNCEME) has the responsibility for advising the government on drought-related decisions. Short-term climate predictions from abroad and local monitoring information in meteorology, hydrology, and agriculture are used to guide state decisions on agricultural and water conservation.

As in Peru, climate-forecast information has helped to alleviate the drastic consequences of ENSO in Brazil (Moura 1994). For example, the normal annual grain production in the state of Ceara is 650 000 tonnes, mostly grown during the rainy season from February to April. In 1987, even though an ENSO forecast had been made (Cane *et al.* 1986), policy makers were unaware of it and no action was taken. The precipitation was 30 percent below normal that year and the total crop production was only 100 000 tonnes. In 1992, the environmental conditions were as poor as at the time of the previous ENSO warming, but the response of the agricultural sector was markedly different. Precipitation for 1992 was 27 percent below the long-term mean, yet the grain production for the year was 530 000 tonnes.

The primary reason for the dramatic improvement was that in late 1991, FUNCEME acted on a combination of ENSO forecasts and information on conditions in the Atlantic, and alerted the government to an impending dry growing season. Then-Governor of Ceara, Ciro Gomes, traveled to the interior of the state with the Secretary of Agriculture and the President of FUNCEME to initiate a series of actions that culminated in the farmers planting crops appropriate for a shorter growing season. Potentially catastrophic conditions were averted for the general populace when FUNCEME combined observations of the state's water resources with the drought forecast, and advised that the capital city of Fortaleza would run out of drinking water by December 1992. The state acted to restrict water consumption and invested U.S.$13 million to construct a new dam. In 1993, actions were taken in anticipation of a second year of drought. Rainfall was 45 percent below the norm and grain production dropped to 250 000 tonnes, yet this level was still 2.5 times more than the total production in 1987. The responses encouraged by FUNCEME prevented what could have been the disastrous consequences of a second consecutive year of drought.

Australia

South America is not the only continent where short-term climate predictions have been put to regular use. The seasonal rainfall of Australia is characterized by significant year-to-year variability, much of which is linked to ENSO. Simpson et al. (1993) showed that annual flow variations for the River Murray, Australia's most extensive river system, are negatively correlated with sea surface temperature in the eastern equatorial Pacific. Encephalitis epidemics are known to occur in the Murray Valley during periods of high rainfall and flooding that coincide with cold phases of ENSO (Nicholls 1994). Because of the severe impact of ENSO-diminished rainfall on rural agricultural production, the Australian Bureau of Meteorology has, since 1988, been issuing seasonal climate outlooks. The primary basis for the outlooks is the status and forecast of ENSO. Information is disseminated by the national and state governments, and through the media. Going into the 1991–1992 ENSO warming, rainfall over much of Queensland and northwestern Australia was generally below to very-much-below average for the period November 1991 to January 1992. Warnings of the forthcoming shortfall in precipitation were issued up to three months in advance, permitting planning by the Queensland Department of Primary Industries. The use of the climate outlooks has tended to flatten variations in crop yields from year to year as a result of mitigating actions by farmers. The success of these outlooks has prompted the Queensland State Government to establish an impacts, monitoring, and decision support system to provide advice at the regional and even at the individual-farm level.

Asia

In Asia, experimental ENSO forecasts are a key input to the monsoon-forecast schemes of the Indian Weather Service. The relationship between ENSO and monsoon variability is indisputable (Webster and Yang 1992), but still an active research topic. In Japan, an ENSO-prediction division has been established within the Japanese Meteorological Agency (JMA). ENSO forecasts based on the predicted displacement of the jet stream are used to develop outlooks for summer and winter temperatures. JMA successfully issued long-range forecasts of a cool and less sunny 1991 boreal summer and a mild 1991–1992 winter, on the basis of a May 1991 prediction of the 1991–1992 warming. In the Philippines, a drought early-warning and monitoring system has been established. Seasonal rainfall advisories are based primarily on ENSO conditions.

Other Foreign Regions

In addition to the examples presented above, where ENSO forecasts are beginning to be used on a routine basis, TOGA research has pointed to many other regions and sectors that could benefit from systematic ENSO forecasts. The National Meteorological Service of Ethiopia has been issuing experimental seasonal forecasts based on ENSO predictions since 1987 (Bekele 1994). Advisories are released for the spring and fall growing seasons. The recommendations take into account land-use strategy, conservation policies, and economic assistance policy. The development of a seasonal forecast model has been hampered by a shortage of Ethiopian personnel trained in long-range forecasting.

Southern Africa was one of the regions dramatically affected by climate variations during 1991–1992. The drought there was worse than any other in that region for the past hundred years. By December 1991, rainfall accumulation was still less than half of the long-term mean throughout Mozambique and southern Zimbabwe. By January 1992, precipitation was as much as 70 percent below normal throughout the southern portion of the continent. Nearly 100 million people were affected, and 12 million tonnes of commodities were imported by relief organizations. Cane *et al.* (1994) showed that there is a strong link between ENSO and interannual fluctuations of both precipitation and maize production in Zimbabwe. They also demonstrated that coupled-model predictions of ENSO could be used to provide an accurate forecast of the

Zimbabwe maize yield, with lead times up to one year. This information could be a valuable component of famine early-warning systems for the region.

Rosenzweig (1995) provided a global perspective on the effects of ENSO on crop yields. Her preliminary work indicated that yields of five major crops in about thirty countries appear to be correlated with ENSO variations. For most of the crops, yields tended to be lower in years with anomalously warm equatorial Pacific sea surface temperature. One-third of the world's production of maize, wheat, sorghum, soybeans, and rice was found to be correlated with phases of ENSO, although for many regions the signal was small.

United States

Investigations of links between agricultural production and ENSO are just beginning for the United States. Adams *et al.* (1995) used the historical temperature and precipitation data for the southeastern United States, together with a model of crop yield, to estimate that climate variations associated with ENSO could decrease crop yield in the southeast by up to 15 percent. The value of improving an ENSO forecast from correlations at the level of roughly 0.6 to a level of 0.8 at lead times of six months in advance of planting decisions was estimated to be more than $100 million per year for the southern agricultural sector alone.

Skill for predictions outside the tropics is likely to be much lower than skill for the tropics because of the greater variability of the climate around the mean. Nevertheless, experimental prediction activities for the United States have begun (Ji *et al.* 1994a, Kerr 1994). Near the end of TOGA, the National Meteorological Center began to use coupled ocean–atmosphere forecasts, together with several statistical techniques, to arrive at an official consensus forecast with lead times up to two seasons in advance.

7. THE FUTURE

TOGA accomplished much in its decade, especially in observing, understanding and predicting ENSO in the tropical Pacific. However, many questions about ENSO and other types of interannual variability around the globe, especially in the middle latitudes, remain unanswered. We recommend for the future: maintenance of observing systems; creation of an institute for developing applications of short-term climate forecasts; and a program for continued research on seasonal-to-interannual climate variations and their predictability.

TOGA went far towards fulfilling its goals—perhaps further than anyone could have foreseen at the beginning of the program. There were many successes. In particular, researchers associated with TOGA:

- conducted process experiments, especially the TOGA Coupled Ocean–Atmosphere Response Experiment (COARE), and processed, distributed, and archived the resulting data sets;
- built and maintained the TOGA Observing System and developed the new technology involved in TOGA Tropical Atmosphere/Ocean (TAO) array of moorings;
- made the data from the TOGA Observing System widely available, in real time, through electronic and paper distribution;
- developed theories for El Niño and the Southern Oscillation (ENSO), using coupled models;
- developed methods of ENSO prediction and demonstrated predictive skill through the establishment and maintenance of the TOGA Program on Prediction (T-POP) and the establishment of the coupled-modeling project at the National Meteorological Center;
- developed effective management and advisory structures to maintain and support the TOGA effort, both nationally and internationally; and
- planned for an International Research Institute for Climate Prediction.

All these accomplishments contributed to a successful program, well balanced and integrated among theory, modeling and predicting, and observations.

TOGA was able to achieve these successes in part because it concentrated on the strongest interannual climate variation: ENSO in the Pacific. This concentration came about for both scientific and programmatic reasons. Scientifically, the engagement of the challenges in the Pacific proved so fruitful that once these were addressed, the TOGA community maintained its enthusiasm as oceanographic, atmospheric, and coupled-system problems were attacked. Funding and support were always only barely adequate, so focusing the resources on a limited area of the Pacific became a necessity.

WHAT TOGA DIDN'T DO

TOGA's limited resources were focused on studying interannual variability in the tropical Pacific associated with ENSO. Consequently, other problems of importance, either ones known at the beginning of TOGA or issues that were identified during the program, remained unexplored. Although the focus on ENSO was consistent with the initial U.S. plan for TOGA (NRC 1983), the program did not cover the broader initial objectives for the international program (WCRP 1985), which were agreed on by U.S. scientists (NRC 1986).

It is well known from empirical studies that the global impacts of ENSO are strongly controlled by the annual cycle, and, in fact, can be thought of as modulations of the mean annual cycle. The very concept of anomalies requires the annual cycle to be known. While the climatology for sea surface temperature is well characterized, the climatologies of other basic quantities in the tropical Pacific—thermocline depth, surface currents, subsurface currents, and subsurface temperature—are still largely unknown, and will be determined only through many years of measurements. Furthermore, the annual cycle itself has components that result from strong coupling among ocean, atmosphere, and land processes. The skill of predictions of sea surface temperature in the tropical Pacific has proven to be strongly dependent on which seasons lie between the time at which and the time for which a prediction is made. Hence, further study is needed on the nature of the annual cycle and its strong variations around the globe, the impact of the annual cycle on interannual variability, and the predictability of interactions between the annual cycle and interannual variations.

Even the nature of ENSO, the basic interannual variability in the tropical Pacific, was not completely determined by TOGA. A ten-year program simply was not long enough to define the nature of ENSO's interannual variability. The nature of interannual variability outside the tropical Pacific remains also underexplored. For example, interannual variability of the strength and onset of the Asian monsoon and the midlatitude connections to ENSO present many unsolved problems. Much remains unknown about seasonal-to-interannual variability induced by midlatitude interactions on large scales between the atmosphere and oceans; induced by the interactions of the atmosphere with sea ice, snow, and land; and induced by the high-frequency forcing by the atmosphere on each of these slow reservoirs (the so-called Hasselmann (1976) mechanism). In considering these issues, it will undoubtedly prove true that the radiative effects of natural and anthropogenic aerosols in producing seasonal-to-interannual variability, and in affecting such variability produced by other physical mechanisms, will have to be considered (see NRC 1996).

Interactions between events in the tropics and extratropics, and the possibility of predicting midlatitude climatic variations, provided some of the justifications for the TOGA Program. However, some issues did not receive the attention they deserved from the TOGA Program. These issues include the role of tropical sea-surface-temperature anomalies in perturbing the extratropical atmosphere, the generation of midlatitude sea-surface-temperature anomalies through this interaction, and the regions of the globe where skillful forecasts of tropical sea surface temperature can be translated into useful regional forecasts. They have been addressed to some degree by other independent efforts, but still need significant attention.

We have come to understand that the warm phase of ENSO can be identified with the eastward extension into the central Pacific of a significant portion of the precipitation normally found in the far western Pacific (the "maritime-continent heat source"). The global circulations arising from the thermal forcing from this gigantic heat source influence other tropical heat sources, in particular the South Pacific Convergence Zone, the Inter-Tropical Convergence Zones, and the heat source over South America. These in turn influence the maritime-continent heat source. Their interactions propagate ENSO influences throughout the entire tropics. A basic question that needs to be addressed is: What influences the position, strength, and interactions among all of the tropical heat sources on monthly to interannual time scales? The interactions can take many forms. The Asian monsoon interacts with ENSO. Interactions of ENSO with the South American heat source influence South American rainfall. ENSO affects the Inter-Tropical Convergence Zone in the Atlantic and, consequently, rainfall in northeast Brazil. ENSO also interacts with the Inter-Tropical Convergence Zone in the eastern Pacific, with attendant interactions between the atmosphere and surface conditions over South America, Mexico, and the United States. All these interactions, which occur relatively slowly, are influenced by the patterns of sea surface temperature and by land processes. The relative slowness of the evolution of the system offers the possibility of being able to predict sea surface temperature using coupled atmosphere–ocean models. In turn, sea surface temperature is a boundary condition for models in which the interactions between the land surface and atmosphere are more accurately simulated, e.g., using nested-grid high-resolution regional models.

It has become clear that there are other modes of interannual variation besides ENSO. Land wetness induces interannual variations of weather, independent of connections to ENSO (see, e.g., Delworth and Manabe 1989). The middle-latitude ocean, atmosphere, and land system would have interannual variability in the absence of ENSO (Manabe and Hahn 1981, N.-C. Lau 1981). Whether these variations are predictable and whether they are enhanced by signals from the tropics remain crucial questions. In addition, the interactions of higher-frequency fluctuations in the tropics and middle latitudes with seasonal-to-interannual climate variations, and the predictability of the resulting climate variations, need to be assessed. Shortly before TOGA started, a clear picture of the "canonical" evolution of ENSO (Rasmusson and Carpenter 1982) was generally believed. However, after the 1982–83 El Niño, the TOGA decade began with a knowledge that the Rasmusson and Carpenter description could not be correct. As TOGA ends, the picture is still not clear. From 1990 to 1994, the climate system appeared to be locked in an unusually extended warm state. The NINO4 region (160°E to 150°W, near the equator) showed persistent positive (0.5–1.0°C) anomalies of sea surface temperature, and the Southern Oscillation Index was negative for that entire period. Such conditions are unprecedented in the instrumental record (Trenberth and Hoar 1996). Is this the result of interactions between interannual and interdecadal variability, or is it a manifestation of global warming? Clearly the nature of these interactions needs to be better understood and quantified.

The key processes that are essential in the coupling of the ocean and atmosphere on seasonal-to-interannual time scales need to be better identified and parameterized in models. The COARE field program was designed to address some of these processes, but was conducted so late in the TOGA Program that it had little influence on prediction and modeling during TOGA. COARE analysis is scheduled to proceed for several years beyond the TOGA time frame, and it is likely that improvements in cloud, surface flux, and boundary layer parameterizations will be made. These studies no doubt will point to additional improvements needed in the convective and mixing parameterizations of coupled ocean–atmosphere–land models. As models become more comprehensive, and as practical forecasting experience is gained, the observing-system requirements for initializing the forecast system will require refinement and re-evaluation. The crucial upper-ocean and land-surface variables requiring initialization need to be identified, and routine observations of these need to be implemented.

OBSTACLES TO PROGRESS

The development of short-term climate prediction and its supporting research enterprise is fraught with difficulties. The cooperation and collaboration of diverse communities is required because ENSO is a coupled atmosphere–ocean phenomenon. Progress in climate prediction will depend on additional coupling in models of the atmosphere and ocean to the land and ice. Development of this complex enterprise will put strain on the existing institutional structures.

Compartmentalization

Traditional departmental structures on college campuses serve to maintain excellence in basic scientific disciplines (e.g., physics, chemistry, biology, geology, and mathematics). However, these structures can discourage study and research on climate problems, which cross disciplinary boundaries. Only the exceptional student can overcome the lack of flexibility in course requirements to develop a path for multidisciplinary climate studies. Only an exceptional faculty member can overcome a lack of institutional rewards for collaborative work.

Some parts of the federal funding structure are organized around traditional disciplinary lines. This organization makes it difficult to plan cross-disciplinary ventures. The investment required for a serious effort to predict seasonal-to-interannual climate variability by means of quantitative physical models is modest in view of the potential economic benefits of skillful forecasts.

There has been a lack of communication between scientists and decision makers in both the public and the private sectors. Most scientists lack knowledge of the day-to-day requirements of decision makers in climate-sensitive portions of the economy. Most decision makers believe that scientists are incapable of understanding and responding in a significant way to needs in the public and private sector.

Building and Maintaining the Infrastructure

The systematic development of a climate-observing system requires long-range planning and commitment by both scientists and governments. The social and political environment evolves on time scales shorter than many climate variations. The geographic coverage of the present *in situ* climate observation network continues to decline at a time when new initiatives are needed (NRC 1994a).

Under stringent budgetary conditions, the building of infrastructure always suffers. A new area needs facilities and resources that are dedicated to its own problems. For short-term climate prediction, these include educational programs, supercomputers, facilities for maintaining the observations needed to initialize and evaluate models, and funding structures that recognize the need to support and coordinate both development and research.

TOGA implemented a prototypical observing and prediction system for limited aspects of ENSO. Such an extensive *in situ* observing system was not foreseen at the beginning of TOGA. As a result, the TAO part of the observing system came into existence very late in the program, almost as TOGA was ending. Thus, while TOGA designed and put an observing system into place, there was no time within the life of the program for evaluation of that system. There is a need to see what effects the system's meteorological observations have on the global forecasts produced by the world's (atmospheric) operational forecast units or what effects its oceanographic observations have on the regular and systematic production of forecasts of sea surface temperature in the tropical Pacific. Furthermore, recent analyses of the decadal variability of tropical sea surface temperature indicate that the observing system may have been designed with too narrow a spatial extent, and may neglect regions important for the understanding of ENSO and its connections to other parts of the globe.

Much remains to be done for implementing a comprehensive system for seasonal-to-interannual climate prediction. The upper-air observing system seems to be decaying. The status of operational satellites is always in question, primarily because of the large resources required. Research satellites do not seem to have the continuity required for obtaining the long, homogenous records needed for the study of climate variations. The effects of the decline of the global observing system on TOGA was discussed in NRC 1992. While outside the direct control of the TOGA community, these problems existed for the duration of the TOGA Program and affected its ability to improve climate understanding and prediction. For future progress in the study of climate variations, it is essential to maintain what we already have, including the upper-air observing network, satellite altimetry, and the upper-ocean and surface-meteorological measurements made routinely in and over the ocean. We re-emphasize the main conclusions of NRC 1994a:

> **Present TOGA observations should be continued. The single most critical effort to be sustained, because of its late establishment and because of the central importance in TOGA predictions of the fields it measures (tropical wind stress, sea surface temperature, upper-ocean thermal structure), is the full TOGA TAO array of approximately 70 moorings."** Furthermore, "Observations now in place to support prediction ... or added later for this reason ... should thereafter be sustained until such time as a serious study of their impact on predictions reveals them to be of marginal value or until a new and more cost-effective technique is demonstrably ready to replace them without disrupting or biasing the geophysical time series."

AN INTERNATIONAL RESEARCH INSTITUTE FOR CLIMATE PREDICTION (IRICP)

In response to the applications being developed for the countries of the tropical Pacific, the Intergovernmental TOGA Board urged the United States to take the lead in designing an institute for experimental short-range climate prediction. This institute would employ the scientific knowledge being developed about ENSO and work with the most affected countries to learn how to use these forecasts. The initially proposed goals for the International Research Institute for Climate Prediction (IRICP) (Moura 1992) are:
1. to continually develop dynamically and thermodynamically consistent coupled models of the global atmosphere, ocean, and land, to serve as a basis for improved climate prediction;
2. to systematically explore the predictability of climate anomalies on time scales out to a few years;
3. to receive, analyze, and archive global atmospheric and oceanic data to improve the scope and accuracy of the forecasts;
4. to systematically produce useful climate forecasts on time scales of several months to several years on global space scales; and
5. to shape and augment these forecasts by incorporating additional physical, agricultural, economic, and other appropriate data, to the explicit social and economic benefit of national societies.

An international meeting to codify the institute's mission and architecture was held during November 1995 in Washington, D.C. This meeting was a major step toward the institute's formal establishment. The intention is that the IRICP provide an international structure that combines worldwide research capabilities, operations, and applications expertise. Regional application centers would certainly play a central role in the interpretation and dissemination of forecasts tailored to local needs. The mission-oriented focus of the IRICP, will help realize the goals of improved predictions and the development of applications of those predictions, with obvious benefit to societies worldwide.

In whatever final form it takes, an IRICP will surely become one of the most important legacies of the TOGA program. TOGA has shown that a well conceived and well executed research program can play a central role in both basic research and policy applications. The application of ENSO predictions for the benefit of the societies and economies of the tropical countries affected by ENSO has taken TOGA out of the realm of pure research. As a corollary, seasonal-to-interannual climate research can justify additional demands on the public treasury.

The IRICP will be a multinational facility with strong connections to the international research community and to application centers in member nations.

It is designed to make the best possible research forecasts of global seasonal-to-interannual climate variations, initially in and around the tropical Pacific, and to distribute those forecasts to applications centers, interested researchers, and operational weather-prediction centers. Training of people from member nations in the nature and use of prediction systems, and tailoring forecasts for specific region, will be central tasks.

The recent Seasonal-to-Interannual Climate Prediction Program (SCPP) proposal (NOAA 1994), as forwarded by NOAA in partnership with other U.S. agencies, is an evolution of the concepts expressed in the IRICP proposal. It is a broad vision of how the United States, both nationally through the National Centers for Environmental Prediction (NCEP, McPherson 1994) and in a multinational framework through the IRICP, would implement these concepts. The SCPP provides mechanisms for the next steps in developing experimental forecasts, in research, in observations, in data management, and in applications of climate information on seasonal-to-interannual time scales. Laying the foundations for a program that will systematically produce and disseminate climate forecasts for applications in societal and economic planning is central to the SCPP. Support for an institute to develop applications of climate prediction can be found in NRC 1994b and 1995b. We re-emphasize a recommendation of NRC 1995b:

> Establish an *international research prototype prediction capability,* including a focused facility (the proposed International Research Institute) and a supporting research program in order to
> - accelerate the application of demonstrated prediction capabilities;
> - secure multinational support for global scale observing systems and international research programs; and
> - focus research to extend predictive capabilities and applications.

GOALS AND CLIVAR

Comprehensive coupled ocean–atmosphere–land models, and prediction systems based on such models, are in their infancy. Development of these models and systems require process studies and the maintenance of appropriate observing systems. The questions raised during TOGA point to the need to develop more comprehensive prediction systems and expand the scope of climate prediction to the entire globe and to longer time scales. The new international program for A Study of Climate Variability and Predictability, CLIVAR (see

WCRP 1995), especially its seasonal-to-interannual component, the Global Ocean–Atmosphere–Land System (GOALS, see WCRP 1995 and NRC 1994b), is designed to address these problems. We re-emphasize the words in NRC 1994b:

> **The ultimate scientific objectives of the GOALS Program would be to:**
> - understand global climate variability on seasonal-to-interannual time scales;
> - determine the spatial and temporal extent to which this variability is predictable;
> - develop the observational, theoretical, and computational means to predict this variability; and
> - make enhanced climate predictions on seasonal-to-interannual time scales.
>
> **The focus of the GOALS program is an assessment of the global interannual climate variation that can be understood, simulated and predicted....**
>
> **It is proposed that the GOALS program be an important component of the Climate Variability and Predictability (CLIVAR) program, which is a broader new initiative of the World Climate Research Program[me] (WCRP) addressing the variability and predictability of the coupled climate system.**

TOGA opened the way to the future of seasonal-to-interannual climate prediction. These follow-on programs will further develop the means of predicting the climate for the ultimate benefit of humankind.

REFERENCES

Adams, R.M., K.J. Bryant, B.A. McCarl, D.M. Legler, J. O'Brien, and A. Solow 1995. The value of improved ENSO forecasts: An example from U.S. agriculture. *Contemporary Economic Policy* **13**, 10–19.

Anderson, D.L.T. and P.B. Rowlands 1976. The role of inertia-gravity and planetary waves in the response of a tropical ocean to the incidence of an equatorial Kelvin wave on a meridional boundary. *J. Mar. Res.* **34**, 295–312.

Anderson, S.P., R.A. Weller, and R. Lukas 1996. Surface buoyancy forcing and the mixed layer of the western Pacific warm pool: Observations and 1-D model results. *J. Climate* **9**, 3056–3085.

Atlas, R., S.C. Bloom, R.N. Hoffman, J. Ardizzone, and G. Brin 1991. Space-based surface wind vectors to aid understanding of air–sea interactions. *EOS* **72**, 201–208.

Bacastow, R.B. 1976. Modulation of atmospheric CO_2 by the Southern Oscillation. *Nature* **261**, 116–118.

Balmaseda, M.A., M.K. Davey, and D.L.T. Anderson 1995: Decadal and seasonal dependence of ENSO prediction skill. *J. Climate* **8**, 2705–2715.

Barber, R.T. 1988. The ocean basin ecosystem. In *Concepts of Ecosystem Ecology*, J. J. Alberts and L. R. Pomeroy (eds.), Springer-Verlag, 166–188.

Barber, R.T. and F.P. Chavez 1983. Biological consequences of El Niño. *Science* **222**, 1203–1210.

Barnett, T.P. 1981. Statistical prediction of North American air temperatures from Pacific predictors. *Mon. Weather Rev.* **109**, 1021–1041.

Barnett, T.P. 1984. Interaction of the monsoon and Pacific trade wind system at interannual time scales. Part III: A partial anatomy of the Southern Oscillation. *Mon. Weather Rev.* **112**, 2388–2400.

Barnett, T.P. 1985. Variations in near-global sea level pressure. *J. Atmos. Sci.* **42**, 478–500.

Barnett, T.P., N.E. Graham, M.A. Cane, S.E. Zebiak, S.C. Dolan, J.J. O'Brien, and D.M. Legler 1988. On the prediction of the El Niño of 1986–1987. *Science* **241**, 192–196.

Barnett, T.P., L. Dumenil, U. Schlese, E. Roeckler, and M. Latif 1989. The effect of Eurasian snow cover on regional and global climate variations. *J. Atmos. Sci.* **46**, 661–685.

Barnett, T.P., M. Latif, E. Kirk, and E. Roeckner 1991. On ENSO physics. *J. Climate* **4**, 487–515.

Barnston, A.G. and C.F. Ropelewski 1992. Prediction of ENSO episodes using canonical correlation analysis. *J. Climate* **5**, 1316–1345.

Barnston, A.G., H.M. van Dool, S.E. Zebiak, T.P. Barnett, M. Ji, D.R. Rodenhuis, M.A. Cane, A. Leetmaa, N.E. Graham, C.R. Ropelewski, V.E. Kousky, E.O. O'Lenic, and R.E. Livesey 1994. Long-lead seasonal forecasts—Where do we stand? *Bull. Am. Meteorol. Soc.* **75**, 2097–2114.

Battisti, D.S. 1988. Dynamics and thermodynamics of a warming event in a coupled tropical atmosphere–ocean model. *J. Atmos. Sci.* **45**, 2889–2919.

Battisti, D.S. and A.C. Hirst 1989. Interannual variability in a tropical atmosphere–ocean model: Influence of the basic state, ocean geometry and nonlinearity. *J. Atmos. Sci.* **46**, 1687–1712.

Battisti, D.S. and E.S. Sarachik 1995. Understanding and predicting ENSO. *Rev. Geophys.* **33**, 1367–1376.

Bekele, F. 1994. Ethiopian use of ENSO information in its seasonal forecasts. In M. Glantz (ed.) *Usable Science: Food Security, Early Warning and El Niño*, National Center for Atmospheric Research, Boulder, Colorado, 117–121.

Bjerknes, J. 1966. A possible response of the atmospheric Hadley circulation to anomalies of ocean temperature. *Tellus* **18**, 820–828.

Bjerknes, J. 1969. Atmospheric teleconnections from the equatorial Pacific. *Mon. Weather Rev.* **97**, 163–172.

Blackmon, M.L., J.E. Geisler, and E. Pitcher 1983. A general-circulation model study of January climate anomaly patterns associated with internnual variation of equatorial Pacific sea surface temperatures. *J. Atmos. Sci.* **40**, 1410–1425.

Blumenthal, M.B. 1991. Predictability of a coupled atmosphere–ocean model. *J. Climate* **4**, 766–784.

Boer, G.J. 1985. Modelling the atmospheric response to the 1982/83 El Niño. In Nihoul 1985, 7–27.

Bradley, F. and R. Weller (eds.) 1995a. Joint Workshop of the TOGA COARE Flux and Atmospheric Working Groups, Boulder, Colorado, 11–13 July. TOGA COARE International Project Office, University Corporation for Atmospheric Research, Boulder, Colorado, 35 pp.

Bradley, F. and R. Weller (eds.) 1995b. Third Workshop of the TOGA COARE Air–Sea Interaction (Flux) Working Group, Honolulu, Hawaii, 2–4 August 1995. TOGA COARE International Project Office, University Corporation for Atmospheric Research, Boulder, Colorado, 34 pp.

Branstator, G.W. 1985. Analysis of general-circulation model sea-surface temperature anomaly sumulations using a linear model. Part I: Forced solutions. *J. Atmos. Sci.* **42**, 2225–2241.

Branstator, G.W. 1995. Organization of stormtrack anomalies by recurring low frequency circulation anomalies. *J. Atmos. Sci.* **52**, 207–226.

Brown, W.S., W.E. Johns, K.D. Leaman, J.P. McCreary, R.L. Molinari, P.L. Richardson, and C. Rooth 1992. A Western Tropical Atlantic Experiment (WESTRAX). *Oceanography* **5**, 73–77.
Busalacchi, A.J. and J.J. O'Brien 1980. The seasonal variability in a model of the tropical Pacific. *J. Phys. Oceanogr.* **10**, 1929–1951.
Busalacchi, A.J. and J.J. O'Brien 1981. Interannual variability of the equatorial Pacific in the 1960's. *J. Geophys. Res.* **86**, 10901–10907.
Busalacchi, A.J., R.M. Atlas, and E.C. Hackert 1993. Comparison of Special Sensor Microwave Imager vector wind stress with model-derived and subjective products for the tropical Pacific. *J. Geophys. Res.* **98**, 6961–6977.
Busalacchi, A.J., M.J. McPhaden, and J. Picaut 1994. Variability in equatorial Pacific sea surface topography during the verification phase of the TOPEX/POSEIDON mission. *J. Geophys. Res.* **99** (C12), 24725–24738.
Canby, T. 1984. El Niño's ill wind. *Natl. Geogr.* **165**, 144–183.
Cane, M.A. and E.S. Sarachik 1977. Forced baroclinic ocean motions. II. The linear equatorial bounded case. *J. Mar. Res.* **35**, 395–432.
Cane, M.A. and E.S. Sarachik 1978. Forced baroclinic ocean motions. III. The equatorial basin case. *J. Mar. Res.* **37**, 355–398.
Cane, M.A. and D.W. Moore 1981. A note on low-frequency equatorial basin modes. *J. Phys. Oceanogr.* **11**, 1578–1585.
Cane, M.A. 1983. Oceanographic events during El Niño. *Science* **222**, 1189–1195.
Cane, M. A. 1984. Modeling sea level during El Niño. *J. Phys. Oceanogr.* **14**, 1864–1874.
Cane, M.A. and P.R. Gent 1984. Reflection of low frequency equatorial waves at arbitrary western boundaries. *J. Mar. Res.* **42**, 487–502.
Cane, M.A., S.E. Zebiak, and S.C. Dolan 1986. Experimental forecasts of El Niño. *Nature* **321**, 827–832.
Cane, M.A., M. Münnich, and S.E. Zebiak 1990. A study of self-excited oscillations of the tropical ocean atmosphere system. Part I: Linear analysis. *J. Atmos. Sci.* **47**, 1562–1577.
Cane, M.A. and E.S. Sarachik, co-chairpersons of the Provisional Working Group 1991. *Prospectus: A TOGA Program on Seasonal-to-Interannual Prediction*. NOAA Climate and Global Change Program Special Report No. 4, University Corporation for Atmospheric Research, 46pp.
Cane, M.A., G. Eshel, and R.W. Buckland 1994. Forecasting Zimbabwean maize yield using eastern equatorial Pacific sea surface temperature. *Nature* **370**, 204–205.
CEES (Committee on Earth and Environmental Sciences of the Federal Coordinating Council for Science, Engineering and Technology) 1993. *Our Changing Planet: The FY94 U.S. Global Change Research Program*. Executive Office of the President, Washington, D.C., 84 pp.

CEPEX [no date]. Central Equatorial Pacific Experiment: Experimental Design. Center for Clouds, Chemistry, and Climate, Scripps Institution of Oceanography, La Jolla, California, 56 pp.

Chang, P. 1994. A study of the seasonal cycle of sea-surface temperature in the tropical Pacific Ocean using reduced gravity models. *J. Geophys. Res.* **99**, 7725–7741.

Chang, P., B. Wang, T. Li, and L. Ji 1994. Interactions between the seasonal cycle and ENSO—Frequency, entrainment, and chaos in a coupled atmosphere–ocean model. *Geophys. Res. Lett.* **21**, 2817–2820.

Charney, J.G., R.G. Fleagle, H. Riehl, V.E. Lally, and D.Q. Wark 1966. The feasibility of a global observation and analysis experiment. *Bull. Am. Phys. Soc.* **47**, 200–220.

Charney, J.G. and J. Shukla 1981. Predictability of monsoons. In *Monsoon Dynamics,* Proceedings of the Joint IUTAM/IUGG International Symposium on Monsoon Dynamics, New Delhi, India, 5–9 December 1977, J. Lighthill and R.P. Pierce (eds.), Cambridge University Press, 99–110.

Chen, D., S.E. Zebiak, A.J. Busalacchi, and M.A. Cane 1995. An improved procedure for El Niño forecasting: Implications for predictability. *Science* **269**, 1699–1702.

Chen, S.S., R.A. Houze, Jr., and B.E. Mapes 1996. Multiscale variability of deep convection in relation to large-scale circulation in TOGA COARE. *J. Atmos Sci.* **53**, 1380–1409.

Cheney, R.E. and L. Miller 1988. Mapping the 1986–1987 El Niño with GEOSAT altimeter data. *EOS* **69**, 754–755.

Chereskin, T., P.P. Niiler, and W.J. Schmitz 1989. A numerical study of the effects of upper ocean shear on flexible drogue drifters. *J. Atmos. Ocean. Technol.* **6** (2), 243–253.

Chiswell, S.M., K.A. Donohue, and M. Wimbush 1995. Variability in the central equatorial Pacific, 1985–1989. *J. Geophys. Res.* **100**, 15849–15863.

Clarke, A.J. 1983. The reflection of equatorial waves from oceanic boundaries. *J. Phys. Oceanogr.* **13**, 1193–1207.

Clarke, A.J. and B. Li 1994. On the timing of warm and cold El-Niño/Southern-Oscillation events. *J. Climate* **8**, *257–2574.*

Climate Analysis Center 1992. *Climate Diagnostics Bulletin,* January 1992, Climate Analysis Center, U.S. Department of Commerce, Camp Springs, Maryland.

Cole, J.E. and R.G. Fairbanks 1990. The Southern Oscillation recorded in the ^{18}O of corals from Tarawa atoll. *Paleoceanography* **5**, 669–683.

Cole, J.E., R.G. Fairbanks, and G.T. Shen 1993. Recent variability in the Southern Oscillation: Isotopic results from a Tarawa Atoll coral. *Science* **260**, 1790–1793.

Colin, C. and B. Bourles 1994. Western boundary currents and transports off French Guiana as inferred from Pegasus observations. *Oceanologica AcTA* **17**, 143–157.

Collins, W.D., F.P.J. Valero, P.J. Flatau, D. Lubin, H. Grassl, and P. Pilewskie 1996. Radiative effects of convection in the tropical Pacific. *J. Geophys. Res.* **101** (D10), 14999–15012.

Cubash, U. 1985. The mean response of the ECMWF global model to the composite El Niño anomaly in extended range prediction experiments. In Nihoul 1985, 329–343.

Delcroix, T., J. Picaut, and G. Eldin 1991. Equatorial Kelvin and Rossby waves evidenced in the Pacific Ocean through GEOSAT sea level and surface current anomalies. *J. Geophys. Res.* **96**, 3249–3262.

Delworth, T. and S. Manabe 1989. The influence of soil wetness on near-surface atmospheric variability. *J. Climate* **2**, 1447–1462.

Didden, N. and F. Schott 1993. Eddies in the North Brazil Current retroflection region observed by GEOSAT altimetry. *J. Geophys. Res.* **98**, 20121–20131.

Donohue, K.A., M. Wimbush, X. Zhu, S.M. Chiswell, R. Lukas, L. Miller, and H.E. Hurlburt 1994. Five years' central Pacific sea level from *in situ* array, satellite altimeter and numerical model. *Atmos.-Ocean* **32**, 495–506.

Enfield, D.B. 1987. The intraseasonal oscillation on eastern Pacific sea levels: How is it forced? *J. Phys. Oceanogr.* **17**, 1860–1876.

Eriksen, C.C. 1982. Equatorial wave vertical modes observed in a western Pacific island array. *J. Phys. Oceanogr.* **12**, 1206–1227.

Eriksen, C.C., M.B. Blumenthal, S.P. Hayes, and P. Ripa 1983. Wind generated equatorial Kelvin waves observed across the Pacific Ocean. *J. Phys. Oceanogr.* **13**, 1622–1640.

Eriksen, C.C. 1985. Moored observations of deep low frequency motions in the central Pacific Ocean: Vertical structure and interpretation as equatorial waves. *J. Phys. Oceanogr.* **15**, 1085–1113.

Fairall, C.W., E.F. Bradley, D.P. Rogers, J.B. Edson, and G.S. Young 1996. Bulk parameterization of air–sea fluxes for Tropical Ocean Global Atmosphere Coupled Ocean–Atmosphere Response Experiment. *J. Geophys. Res* **101** (C2), 3747–3764.

Farrell, B.F. 1990. Small error dynamics and the predictability of atmospheric flows. *J. Atmos. Sci.* **47**, 2409–2416.

Feely, R.A., R.H. Gammon, B.A. Taft, B.A. Pullen, P.E. Waterman, L.S. Conway, J.F. Gendron, and D.P. Weisgarver 1987. Distribution of chemical tracers in the eastern equatorial Pacific during and after the 1982/83 ENSO event. *J. Geophys. Res.* **92**, 6545–6558.

Feely, R.A., R. Wanninkhof, C.A. Cosca, P.P. Murphy, M.F. Lamb, and M.D. Steckley 1995. CO_2 distributions during the 1991–1992 ENSO event. *Deep-Sea Res.* **42**, 365–386.

Fennessy, M.J., L. Marx, and J. Shukla 1985. GCM sensitivity to the 1982–83 equatorial Pacific sea-surface temperature anomalies. In Nihoul 1985, 121–130.

Fine, R.A., R. Lukas, F.M. Bingham, M.J. Warner, and R.H. Gammon 1994. The western equatorial Pacific is a water mass crossroads. *J. Geophys. Res.* **99**, 25063–25080.

Firing, E., R. Lukas, J. Sadler, and K. Wyrtki 1983. Equatorial Undercurrent disappears during 1982–83 El Niño. *Science* **222**, 1121–1123.

Flament, P., S. Kennan, R. Knox, P. Niiler, and R. Bernstein 1996. The three-dimensional structure of an upper ocean vortex in the tropical Pacific Ocean. *Nature* **383**, 610–613.

Francey, R., P.P. Tans, C.E. Allison, I.G. Enting, J.W.C. White, and M. Trolier 1995. Changes in oceanic and terrestrial carbon uptake since 1982. *Nature* **373**, 326–330.

Freilich, M.H. and R.S. Dunbar 1993. A prliminary C-band scatterometer model function for ERS-1 AMI instrument. In *Proc. First ERS-1 Symposium: Space at the Service of our Environment*, Cannes, 4–6 November 1992, European Space Agency, Paris, 79–84.

Fushimi, K. 1987. Variations of carbon dioxide partial pressure in the western North Pacific surface water during the 1982/83 El Niño event. *Tellus* **39B**, 214–227.

Gage, K.S., C.R. Williams, and W.L. Eklund 1994. UHF wind profilers: A new tool for diagnosing convective cloud systems. *Bull. Am. Meteorol. Soc.* **75**, 2289–2294.

Geisler, J.E., M.L. Backmon, G.T. Bates, and S. Muñoz 1985. Sensitivity of January climate response to the magnitude and position of equatorial Pacific sea surface temperature anomalies. *J. Atmos. Sci.* **42**, 1037–1049.

Gill, A.E. 1980. Some simple solutions for heat-induced tropical circulation. *Quart. J. Roy. Meteorol. Soc.* **106**, 447–462.

Gill, A.E. 1982. *Atmosphere–Ocean Dynamics*. Academic Press, New York, 662pp.

Glantz, M.H. and J.D. Thompson (eds.) 1981. *Resource Management and Environmental Uncertainty: Lessons from Coastal Upwelling Fisheries*. J. Wiley, New York.

Goldenberg, S.B. and J.J. O'Brien 1981. Time and space variability of tropical Pacific wind stress. *Mon. Weather Rev.* **109**, 1190–1205.

Goswami, B.N. and J. Shukla 1991. Predictability of a coupled ocean–atmosphere model. *J. Climate* **4**, 107–115.

Graham, N.E., J. Michaelsen, and T.P. Barnett 1987a. An investigation of the El Niño–Southern Oscillation cycle with statistical models. 1. Predictor field characteristics. *J. Geophys. Res.* **92**, 14251–14270.

Graham, N.E., J. Michaelsen, and T.P. Barnett 1987b. An investigation of the El Niño–Southern Oscillation cycle with statistical models. 2. Model results. *J. Geophys. Res.* **92**, 14271–14289.

Graham, N.E., T.P. Barnett, M.A. Cane, and S.E. Zebiak 1994. Simulated greenhouse warming and model ENSO cycles. Scripps Institution of Oceanography Reference Series 94-04, University of California–San Diego, La Jolla, 17pp.

Gurney, R.J., J.L. Foster, and C.L. Parkinson (eds.) 1993. *Atlas of Satellite Observations Related to Global Change.* Cambridge University Press, 470pp.

Gutzler, D.S., G.N. Kiladis, G.A. Meehl, K.M. Weickmann, and M. Wheeler 1994. The global climate of December 1992–February 1993. Part II: Large-scale variability across the tropical Pacific during TOGA COARE. *J. Climate* **7**, 1606–1622.

Halpern, D. 1980. A Pacific equatorial temperature section from 172°E to 110°W during winter–spring 1979. *Deep-Sea Res.* **27**, 931–940.

Halpern, D., S.P. Hayes, A. Leetmaa, D. Hansen, and S.G.H. Philander 1983. Oceanographic observations of the 1982 warming of the tropical eastern Pacific. *Science* **221**, 1173–1175.

Halpern, D. 1987a. Comparison of moored wind measurements from a spar and toroidal buoy in the eastern equatorial Pacific during February–March 1981. *J. Geophys. Res.* **92**, 8303–8306.

Halpern, D. 1987b. Comparison of upper ocean VACM and VMCM observations in the equatorial Pacific. *J. Atmos. Ocean. Technol.* **4**, 84–93.

Halpern, D. 1987c. Observations of annual and El Niño thermal and flow variations along the equator at 0°, 110°W and 0°, 95°W during 1980–1985. *J. Geophys. Res.* **92**, 8197–8212.

Halpern, D. 1988. Moored surface wind observations at four sites along the Pacific equator between 140°W and 95°W. *J. Climate* **1**, 1251–1260.

Halpern, D., R.A. Knox, and D.S. Luther 1988. Observations of 20-day period meridional current oscillations in the upper ocean along the Pacific equator. *J. Phys. Oceanogr.* **18**, 1514–1534.

Halpern, D. and R.H. Weisberg 1989. Upper ocean thermal and flow fields at 0°, 28°W (Atlantic) and 0°, 140°W (Pacific) during 1983–1985. *Deep-Sea Res.* **36**, 407–418.

Halpern, D., H. Ashby, C. Finch, E. Smith, and J. Robles 1990. TOGA CD-ROM description: Meteorological and oceanographic data sets for 1985 and 1986. JPL Publication 90-43, Jet Propulsion Laboratory, Pasadena, California, 43pp.

Halpern, D. 1996. Visiting TOGA's past. *Bull. Am. Meteorol. Soc.* **77**, 233–242.

Hansen, D.V. and C.A. Paul 1984. Genesis and effects of long waves in the equatorial Pacific. *J. Geophys. Res.* **89**, 10431–10440.

Harrison, D.E. 1989. Local and remote forcing of ENSO ocean waveguide response. *J. Phys. Oceanogr.* **19**, 691–699.

Hasselmann, K. 1976. Stochastic climate models. I: Theory. *Tellus* **28**, 473–485.

Hastenrath, S. and L. Heller 1977. Dynamics of climatic hazards in northeast Brazil. *Quart. J. Roy. Meteorol. Soc.* **103**, 77–92.

Hayes, S.P., L.J. Mangum, J. Picaut, A. Sumi, and K. Takeuchi 1991. TOGA-TAO: A moored array for real-time measurements in the tropical Pacific Ocean. *Bull. Am. Meteorol. Soc.* **72**, 339–347.

Heimann, M. 1995. Dynamics of the carbon cycle. *Nature* **375**, 629–630.

Held, I.M., S.W. Lyons, and S. Nigam 1989. Transients and the extratropical response to El Niño. *J. Atmos. Sci.* **46**, 163–174.

Hellerman, S. and M. Rosenstein 1983. Normal monthly wind stress over the world ocean with error estimates. *J. Phys. Oceanogr.* **13**, 1093–1104.

Hirst, A.C. 1986. Unstable and damped equatorial modes in simple coupled ocean–atmosphere models. *J. Atmos. Sci.* **43**, 606–630.

Hirst, A.C. 1988. Slow instabilities in tropical ocean basin–global atmosphere models. *J. Atmos. Sci.* **45**, 830–852.

Hirst, A.C. and K.-M. Lau 1990. Intraseasonal and interannual oscillations in coupled ocean–atmosphere models. *J. Climate* **3**, 713–725.

Holland, G.J., J.L. McBride, R.K. Smith, D. Jasper, and T.D. Keenan 1986. The BMRC Australian Monsoon Experiment: AMEX. *Bull. Am. Meteorol. Soc.* **67**, 1466–1472.

Horel, J.D. and J.M. Wallace 1981. Planetary-scale atmospheric phenomena associated with the Southern Oscillation. *Mon. Weather Rev.* **109**, 813–829.

Hoskins, B.J. and D. Karoly 1981. The steady linear response of a spherical atmosphere to thermal and orographic forcing. *J. Atmos. Sci.* **38**, 1179–1196.

Hurlburt, H.J., J. Kindle, and J.J. O'Brien 1976. A numerical simulation of the onset of El Niño. *J. Phys. Oceanogr.* **6**, 621–631.

IGFA (International Group of Funding Agencies for Global Change Research) 1993. Preliminary Study Report No. 1: IGFA Resource Assessment—1993, available from former IGFA Secretariat, c/o National Science Foundation, Arlington, Virginia.

Inoue, M. and J.J. O'Brien 1984. A forecasting model for the onset of El Niño. *Mon. Weather Rev* **112**, 2326–2337.

Inoue, H.Y. and Y. Sugimura 1992. Variations and distributions of CO_2 in and over the equatorial Pacific during the period from the 1986/87 El Niño to the 1988/89 La Niña event. *Tellus* **44B**, 1–22.

Inoue, H.Y., M. Ishii, H. Matsueda, and M. Ahoyama 1996. Changes in longitudinal distribution of the partial pressure of CO_2 in the central and western Pacific west of 160°W. *Geophys. Res. Lett.* **23**, 1781–1784.

International TOGA Project Office 1992. *TOGA International Implementation Plan* (fourth edition). ITPO Report No. 1, World Meteorological Organization, Geneva, 73pp + appendices.

Ishii, M. and H.Y. Inoue 1995. Air–sea exchange of CO_2 in the central and western equatorial Pacific in 1990. *Tellus* **47B**, 447–460.

Ji, M., A. Kumar, and A. Leetmaa 1994a. A multi-season climate forecast system at the National Meteorological Center. *Bull. Am. Meteorol. Soc.* **75**, 569–577.

Ji, M., A. Kumar, and A. Leetmaa 1994b. An experimental coupled forecast system at the National Meteorological Center: Some early results. *Tellus* **46A**, 398–418.

Jin, F.-F. and D.S. Neelin 1993a. Modes of interannual tropical ocean–atmosphere interaction—A unified view. Part I: Numerical results. *J. Atmos. Sci.* **50**, 3477–3503.

Jin, F.-F. and D.S. Neelin 1993b. Modes of interannual tropical ocean–atmosphere interaction—A unified view. Part III: Analytical results in fully coupled cases. *J. Atmos. Sci.* **50**, 3523–3540.

Jin, F.-F., J.D. Neelin, and M. Ghil 1994. El Niño on the Devil's Staircase: Annual subharmonic steps to chaos. *Science* **264**, 70–72.

Johns, W.E., T.N. Lee, F. Schott, R. Zantopp, and R. Evans 1990. The North Brazil Current retroflection: Seasonal structure and eddy variability. *J. Geophys. Res.* **95**, 22103–22120.

Johnson, R.H., P.E. Ciesielski, and K.A. Hart 1996. Tropical inversions near the 0°C level. *J. Atmos. Sci.* **53**, 1838–1855.

Ju, J. and J.M. Slingo 1995. The Asian summer monsoon and ENSO. *Quart. J. Roy. Meteorol. Soc.* **122**, 1133–1168.

Julian, P.R. and R.M. Chervin 1978. A study of the Southern Oscillation and Walker Circulation phenomenon. *Mon. Weather Rev.* **106**, 1433–1451.

Kawamura, R. 1994. A rotated EOF analysis of global sea surface temperature variability with interannual and interdecadal time scales. *J. Phys. Oceanogr.* **24**, 707–715.

Keeling, C.D. and R. Revelle 1985. Effects of ENSO on the atmospheric content of CO_2. *Meteoritics* **20**, 437–450.

Keeling, C.D., R.B. Bacastow, A.F. Carter, S.C. Piper, T.P. Whorf, M. Heimann, W.G. Mook, and H. Roeloffzen 1989. A three-dimensional model of atmospheric CO_2 transport based on observed winds: Analysis and observational data. In *Aspects of Climate Variability in the Pacific and Western Americas,* Geophysical Monograph **55**, D. Peterson (ed.), American Geophysical Union, Washington, D.C., 165–236.

Keeling, C.D., T.P. Whorf, M. Whalen, and J. Van der Plicht 1995. Interannual extremes in the rate of rise of atmospheric carbon dioxide since 1980. *Nature* **375**, 666–670.Kerr, R.A. 1994. Official forecasts pushed out to a year ahead. *Science* **266**, 1940–1941.

Kerr, R.A. 1994. Official forecasts pushed out to a year ahead. *Science* **266**, 1940–1941.

Keshavamurty, R.N. 1982. Response of the atmosphere to sea surface temperature anomalies over the equatorial Pacific and the teleconnections of the Southern Oscillation. *J. Atmos. Sci.* **39**, 1241–1259.

Kessler, W.S. 1990. Can reflected extra-equatorial Rossby waves drive ENSO? *J. Phys. Oceanogr.* **21**, 444–452.

Klein, S.A. and D.L. Hartmann 1993. The seasonal cycle of low stratiform clouds. *J. Climate* **6**, 1587–1606.

Knox, R.A. and D. Halpern 1982. Long range Kelvin wave propagation of transport variations in Pacific Ocean equatorial currents. *J. Mar. Res.* **40** suppl., 329–339.

Köberle, C. and S.G.H. Philander 1994. On the processes that control seasonal variations of sea surface temperatures in the tropical Pacific Ocean. *Tellus* **46A**, 481–496.

Krishnamurti, T.N., H.S. Bedi, and M. Subramaniam 1989a. The summer monsoon of 1987. *J. Climate* **2**, 321–340.

Krishnamurti, T.N., H.S. Bedi, and M. Subramaniam 1989b. The summer monsoon of 1988. *Meteorol. Atmos. Phys.* **42**, 19–37.

Kumar, A. and M.P. Hoerling 1995. Prospects and limitations of atmospheric GCM climate predictions. *Bull. Am. Meteorol. Soc.* **76**, 335–345.

Kushnir, Y. and N.-C. Lau 1992. The general-circulation model response to a North Pacific sea surface temperature anomaly: Dependence on time scale and pattern polarity. *J. Climate* **5**, 271–283.

Lagos, P. and J. Buizer 1992. El Niño and Peru: A nation's response to interannual climate variability. In *Natural and Technological Disasters: Causes, Effects, and Preventative Measures*, S.K. Majumdar, G.S. Forbes, E.W. Miller, and R.F. Schmaltz (eds.), Pennsylvania Acad. of Science, 223–238.

Latif, M., A. Sterl, E. Maier-Reimer, and M.M. Junge 1993. Climate variability in a coupled GCM. Part I: The tropical Pacific. *J. Climate* **6**, 5–21.

Latif, M., T.P. Barnett, M.A. Cane, M. Flugel, N.E. Graham, H. von Storch, J.-S. Xu, and S.E. Zebiak 1994. A review of ENSO prediction studies. *Climate Dyn.* **9**, 167–179.

Latif, M. and T.P. Barnett 1995. Interactions of the tropical oceans. *J. Climate* **8**, 952–964.

Latif, M., A. Groetzner, M. Munnich, E. Maier-Reimer, S. Venzke, and T.P. Barnett 1996. A mechanism for decadal climate variability. Max Planck Institut für Meteorologie Report No. 187, Hamburg.

Lau, K.-M. and A.J. Busalacchi 1993. El Niño Southern Oscillation: A view from space. In Gurney *et al.* 1993, 281–294.

Lau, N.-C. 1981. A diagnostic study of recurrent meteorological anomalies in a 15-year simulation with a GFDL general circulation model. *Mon. Weather Rev.* **109**, 2287–2311.

Lau, N.-C. and M.J. Nath 1990. A general circulation model study of the atmospheric response to extratropical sea surface temperature anomalies observed in 1950–79. *J. Climate* **3**, 965–989.

Lau, N.-C., S.G.H. Philander, and M.J. Nath 1992. Simulation of El Niño / Southern Oscillation phenomenon with a low-resolution coupled general circulation model of the global ocean and atmosphere. *J. Climate* **5**, 284–307.

Lau, N.-C., and M.J. Nath 1994. A modeling study of the relative roles of tropical and extra-tropical SST anomalies in the variability of the global atmosphere–ocean system. *J. Climate* **7**, 1184–1207.

Lea, D.W., G.T. Shen, and E.A. Boyle 1989. Coralline barium records temporal variability in equatorial Pacific upwelling. *Nature* **340**, 373–376.

Leetmaa, A. and M. Ji 1989. Operational hindcasting of the tropical Pacific. *Dyn. Atmos. Ocean* **13**, 465–490.

Lefevre, N. and Y. Dandonneau 1992. Air–sea CO_2 fluxes in the equatorial Pacific in January–March 1991. *Geophys. Res. Lett.* **19**, 2223–2226.

Legeckis, R. 1977. Long waves in the eastern equatorial Pacific: A view from a geostationary satellite. *Science* **197**, 1171–1181.

Legler, D.M. and J.J. O'Brien 1988. Tropical Pacific wind stress analysis for TOGA. *Time Series of Ocean Measurements*, Technical Series of the International Oceanographic Commission, No. 33, vol. 4, 11–17.

Legler, D.M. 1991. Errors of 5-day mean surface wind and temperature conditions due to inadequate sampling. *J. Atmos. Ocean. Technol.* **8**, 709–712.

Legrand, M. and C. Feniet-Saigne 1991. Methanesulfonic acid in south polar snow layers: A record of strong El Niño? *Geophys. Res. Lett.* **18**, 187–190.

Lewis, M.R., M.E. Carr, G.C. Feldman, W.E. Esaias, and C.R. McClain 1990. Satellite estimates of the influence of penetrating solar radiation on the heat budget of the equatorial Pacific Ocean. *Nature* **347**, 543–545.

Li, B. and A.J. Clarke 1994. An examination of some ENSO mechanisms using interannual sea level at the eastern and western equatorial boundaries and the zonally averaged equatorial wind. *J. Phys. Oceanogr.* **24**, 681–690.

Lin, X. and R.H. Johnson 1996a. Kinematic and thermodynamic characteristics of the flow over the western Pacific warm pool during TOGA COARE. *J. Atmos. Sci.* **53**, 695–715.

Lin, X. and R.H. Johnson 1996b. Heating, moistening, and rainfall over the western Pacific warm pool during TOGA COARE. *J. Atmos. Sci.* **53**, 3367–3383.

Lindstrom, E., R. Lukas, R. Fine, E. Firing, S. Godfrey, G. Meyers, and M. Tsuchiya 1987. The Western Equatorial Pacific Ocean Circulation Study. *Nature* **330**, 533–537.

Liu, W.T. 1986. Statistical relation between monthly precipitable water and surface level humidity over global oceans. *Mon. Weather Rev.* **114**, 1591–1602.

Liu, W.T. 1988. Moisture and latent heat flux variabilities in the tropical Pacific derived from satellite data. *J. Geophys. Res.* **93**, 6749–6760.

van Loon, H. and R.A. Madden 1981. The Southern Oscillation. Part I. Global associations with pressure and temperature in northern winter. *Mon. Weather. Rev.* **109**, 1150–1162.

Lorenz, E.N. 1965. A study of the predictability of a 28-variable atmospheric model. *Tellus* **17**, 321–333.

Lorenz, E.N. 1982. Atmospheric predictability experiments with a large numerical model. *Tellus* **34**, 505–513.

Lukas, R., S. Hayes, and K. Wyrtki 1984. Equatorial sea level response during the 1982–83 El Niño. *J. Geophys. Res.* **89**, 10425–10430.

Lukas, R. 1988. Interannual fluctuations of the Mindanao Current inferred from sea level. *J. Geophys. Res.* **93**, 6744–6748.

Lukas, R. and E. Lindstrom 1991. The mixed layer of the western equatorial Pacific Ocean. *J. Geophys. Res.* **96** suppl., 3343–3357.

Lukas, R., E. Firing, P. Hacker, P.L. Richardson, C.A. Collins, R. Fine, and R. Gammon 1991. Observations of the Mindanao Current during the Western Equatorial Pacific Ocean Circulation Study. *J. Geophys. Res.* **96**, 7089–7104.

Lukas, R., P.J. Webster, and A. Leetmaa 1995. The large-scale context for the TOGA Coupled Ocean–Atmosphere Response Experiment. *Meteorol. Atmos. Phys.* **56**, 3–16.

Luther, D.S. and D.E. Harrison 1984. Observing long-period fluctuations of surface winds in the tropical Pacific: Initial results from island data. *Mon. Weather Rev.* **112**, 285–302.

Magalhaes, A. and P. Magee 1994. The Brazilian Nordeste (Northeast). In *Drought Follows the Plow*, M.H. Glantz (ed.), Cambridge University Press, Cambridge, 197pp.

Manabe, S. and D.G. Hahn 1981. Simulation of atmospheric variability. *Mon. Weather Rev.* **109**, 2260–2286.

Manabe, S. and R.J. Stouffer 1988. Two stable equilibria of a coupled ocean–atmosphere model. *J. Climate* **1**, 841–866.

Mangum, L.J. and S.P. Hayes 1984. The vertical structure of the zonal pressure gradient in the eastern equatorial Pacific. *J. Geophys. Res.* **89**, 10441–10449.

Mann, M.E., and J. Park 1994. Global-scale modes of surface temperature variability on interannual to century time scales. *J. Geophys. Res.* **99**, 25819–25833.

Mantua, N.J. 1994. *Numerical Modeling Studies of the El Niño–Southern Oscillation.* Ph. D. thesis, Univ. of Washington, Seattle, 152pp.

Mantua, N.J. and D.S. Battisti 1994. Evidence for the delayed oscillator mechanism for ENSO: The "observed" oceanic Kelvin mode in the far western Pacific. *J. Phys. Oceanogr.* **24**, 691–699.

Mantua, N.J. and D.S. Battisti 1995. On the role of competing coupled instabilities in the behavior of the Zebiak-Cane and Battisti coupled ocean–atmosphere models. *J. Climate* **8**, 2897–2927.

Mapes, B.E. and R.A. Houze, Jr. 1993. Cloud clusters and superclusters over the oceanic warm pool. *Mon. Weather Rev.* **121**, 1398–1415.

Mass, C.F. and D.A. Portman 1989. Major volcanic eruptions and climate: A critical evaluation. *J. Climate* **2**, 566–593.

Matsuno, T. 1966. Quasi-geostrophic motions in the equatorial area. *J. Meteor. Soc. Japan* **44**, 25–42.

Mayer, D.A. and R.H. Weisberg 1993. A description of COADS surface meteorological fields and the implied Sverdrup transports for the Atlantic Ocean from 30°S to 60°N. *J. Phys. Oceanogr.* **23**, 2201–2221.

McCreary, J.P. 1976. Eastern tropical ocean response to changing wind systems with application to El Niño. *J. Phys. Oceanogr.* **6**, 632–645.

McPhaden, M.J. and A.B. Taft 1988. The dynamics of seasonal and intraseasonal variability in the eastern equatorial Pacific. *J. Phys. Oceanogr.* **18**, 1713–1732.

McPhaden, M.J. and S.P. Hayes 1991. An analysis of monthly period fluctuations in the heat balance of the Pacific North Equatorial Countercurrent. *EOS* **72** suppl., 274.

McPhaden, M.J. and M.E. McCarty 1992. Mean seasonal cycles and interannual variations at 0°, 110° W and 0°, 140°W during 1980–1991. NOAA Technical Memorandum ERL PMEL-95, Seattle, NTIS No. PB93-114726/XAB, 124pp.

McPhaden, M.J. 1996. Monthly period oscillations in the Pacific North Equatorial Countercurrent. *J. Geophys. Res.* **101**, 6337–6359.

McPherson, R.D. 1994. The National Centers for Environmental Prediction: Operational climate, ocean, and weather prediction for the 21st century. *Bull. Am. Meteorol. Soc.* **75**, 363–373.

Mechoso, C.R., A.W. Robertson, N. Barth, M.K. Davey, P. Delecluse, P.R. Gent, S. Ineson, B. Kirtman, M. Latif, H. Le Treut, T. Nagai, J.D. Neelin,

S.G.H. Philander, J. Polcher, P.S. Schopf, T. Stockdale, M.J. Suarez, L. Terray, O. Thual, and J.J. Tribbia 1995. The seasonal cycle over the tropical Pacific in coupled atmosphere–ocean general-circulation models. *Mon. Weather Rev.* **123**, 2825–2838.

Meehl, G.A., G.W. Branstator, and W.M. Washington 1993. Tropical Pacific interannual variability and CO_2 climate change. *J. Climate* **6**, 42–63.

Metzger, E.J., E.H. Hurlburt, J.C. Kindle, Z. Sirkes, and J.M Pringle 1992. Hindcasting of wind-driven anomalies using a reduced gravity global ocean model. *J. Mar. Technol. Soc.* **26**, 23–32.

Miller, L., R.E. Cheney, and B.C. Douglas 1988. GEOSAT altimeter observations of Kelvin waves and the 1986–87 El Niño. *Science* **239**, 52–54.

Mitchum, G. and R. Lukas 1987. The latitude-frequency structure of Pacific sea level variance. *J. Phys. Oceanogr.* **17**, 2362–2365.

Mitchum, G. and R. Lukas 1990. Westward propagation of annual sea level and wind signals in the western Pacific Ocean. *J. Climate* **3**, 1102–1110.

Mitchum, G.T. 1994. Comparison of TOPEX sea surface height and tide gauge sea levels. *J. Geophys. Res.–Oceans* **99**, 24541–24553.

Miyakoda, K., T. Gordon, R. Carerly, W. Stern, J. Sirutis, and W. Bourke 1983. Simulation of a blocking event in January 1977. *Mon. Weather Rev.* **111**, 846–869.

Molinari, R.L. and E. Johns 1994. Upper layer temperature structure of the western tropical Atlantic. *J. Geophys. Res.* **99** (C9), 18225–18233.

Molteni, F., L. Ferranti, T.N. Palmer, and P. Viterbo 1993. A dynamical interpretation of the global response to equatorial Pacific sea surface temperature anomalies. *J. Climate* **6**, 777–795.

Moore, D.W. and S.G.H. Philander 1977. Modeling of the tropical oceanic circulation. In *The Sea,* E.D. Goldberg, J.J. O'Brien, and J.H. Steele (eds.), Wiley Interscience, 319–361.

Moum, J.N., D.R. Caldwell, and C.A. Paulson 1989. Mixing in the equatorial surface layer and thermocline. *J. Geophys. Res.* **94**, 2005–2021.

Moura, A.D. and J. Shukla 1981. On the dynamics of droughts in northeast Brazil: Observations, theory and numerical experiments with a general circulation model. *J. Atmos. Sci.* **38**, 2653–2675.

Moura, A. and the Task Group 1992. International Research Institute for Climate Prediction: A Proposal. Available from the NOAA Office of Global Programs, U.S. Department of Commerce, Silver Spring, Maryland, 51pp.

Moura, A.D. 1994. Prospects for Seasonal-to-Interannual Climate Prediction and Applications for Sustainable Development. *WMO Bulletin* **43**, 207–215.

Munnich, M, M.A. Cane, and S.E. Zebiak 1991. A study of self-excited oscillations of the tropical ocean–atmosphere system. II. Nonlinear cases. *J. Atmos. Sci.* **48**, 1238–1248.

Murray, J.W., M.W. Leinen, R.A. Feeley, Jr., J.R. Toggweiler, and R. Wanninkhof 1992. EqPac: A process study in the Central Equatorial Pacific. *Oceanography* **5**, 134–142.

Murray, J.W., R.T. Barber, M.R. Roman, M.P. Bacon, and R.A. Feeley 1994. Physical and biological controls on carbon cycling in the equatorial Pacific. *Science* **266**, 58–65.

Nagai, T., T. Tokioka, M. Endoh, and Y. Kitamura 1992. El Niño–Southern Oscillation simulated in a MIR atmosphere–ocean coupled general circulation model. *J. Climate* **5**, 1202–1233.

Nagai, T., Y. Kitamura, M. Endoh, and T. Tokioka 1995. Coupled atmosphere–ocean model simulations of El Niño / Southern Oscillation with and without active Indian Ocean. *J. Climate* **8**, 3–14.

Neelin, J.D., M. Latif, M.A.F. Allaart, M.A. Cane, U. Cubasch, W.L. Gates, P.R. Gent, M. Ghil, N.C. Lau, C.R. Mechoso, G.A. Meehl, J.M. Oberhuber, S.G.H. Philander, P.S. Schopf, K.R. Sperber, A. Sterl, T. Tokioka, J. Tribbia, and S.E. Zebiak 1992. Tropical air–sea interaction in general-circulation models. *Clim. Dynam.* **7**, 73–104.

Neelin, J.D. and F.F. Jin 1993. Modes of interannual tropical ocean–atmosphere interaction—A unified view. Part II: Analytical results in the weak coupling limit. *J. Atmos. Sci.* **50**, 3504–3522.

Neelin, J.D., M. Latif, and F.-F Jin 1994. Dynamics of coupled ocean–atmosphere models: The tropical problem. *Ann. Rev. Fluid Mech.* **26**, 617–659.

Newell, R.E. 1979. Climate and the ocean. *Am. Scientist* **67**, 405–416.

Newell, R.E. 1986. An approach towards equilibrium temperature in the tropical eastern Pacific. *J. Phys. Oceanogr.* **16**, 1338–1342.

Nicholls, N. 1994. El Niño / Southern Oscillation and vector-borne disease. *The Lancet* **342**, 1285.

Nihoul, J.C.J. (ed.) 1985. *Coupled Ocean–Atmosphere Models.* Elsevier Oceanography Series, No. 40, papers presented at the 16th International Liege Colloquium on Ocean Hydrodynamics 1984, Elsevier, Amsterdam.

Niiler, P. (ed.) 1982. Tropic Heat: A Study of the Tropical Pacific Upper-Ocean Heat, Mass, and Momentum Budgets—The CORE Research Program. Oregon State University, Corvallis.

Niiler, P.P., R.E. Davis, and H.J. White 1987. Water-following characteristics of a mixed layer drifter. *Deep-Sea Res.* **34**, 1867–1882.

Niiler, P.P., A.S. Sybrandy, K.N. Bi, P.-M. Poulain, and D. Bitterman 1994. Measurements of the water-following capability of holey-sock and TRISTAR drifters. *Deep-Sea Res.* **42**, 1951.

Nitta, T., and S. Yamada 1989. Recent warming of tropical sea surface temperature and its relationship to Northern Hemisphere circulation. *J. Meteorol. Soc. Japan* **67**, 375–383.

NOAA (National Oceanic and Atmospheric Administration) 1994. A Proposal to Launch a Seasonal-to-Interannual Climate Prediction Program. NOAA Office of Global Programs, Silver Spring, Maryland, 19pp.

Normand, C. 1953. Monsoon seasonal forecasting. *Quart. J. Roy. Meteorol. Soc.* **79**, 463–473.

Nova University 1989. *U.S. TOGA Ocean Observing System Mid-Life Progress Review and Recommendations for Continuation: Workshop Report.* Nova University Press, Fort Lauderdale, Florida, 102 pp.

NRC (National Research Council) 1983. *El Niño and the Southern Oscillation: A Scientific Plan.* National Academy Press, Washington, D.C., 72pp.

NRC 1986. *U.S. Participation in the TOGA Program: A Research Strategy.* National Academy Press, Washington, D.C., 24pp.

NRC 1990. *TOGA: A Review of Progress and Future Opportunities.* National Academy Press, Washington, D.C., 66pp.

NRC 1992. *A Decade of International Climate Research: The first ten years of the World Climate Research Program[me].* National Academy Press, Washington, D.C., 59 pp.

NRC 1994a. *Ocean–Atmosphere Observations Supporting Short-Term Climate Predictions.* National Academy Press, Washington, D.C., 51pp.

NRC 1994b. *GOALS (Global Ocean–Atmosphere–Land System) for Predicting Seasonal-to-Interannual Climate.* National Academy Press, Washington, D.C., 103 pp.

NRC 1995a. *Organizing U.S. Participation in GOALS.* National Academy Press, Washington, D.C., 8 pp.

NRC 1995b. *A Review of the U.S. Global Change Research Program and NASA's Mission to Planet Earth / Earth Observing System.* National Academy Press, Washington, 96 pp.

NRC 1996. *Aerosol Radiative Forcing and Climate Change.* National Academy Press, Washington, D.C., 161pp.

O'Brien, J., R. Kirk, L. McGoldrick, J. Witte, R. Atlas, E. Bracalente, O. Brown, R. Haney, D.E. Harrison, D. Honhart, H. Hurlburt, R. Johnson, L. Jones, K. Katsaros, R. Lambertson, S. Peteherych, W. Pierson, J. Price, D. Ross, R. Stewart, and P. Woiceshyn 1982. Scientific opportunities using satellite surface wind stress measurements over the ocean. In *Report of the Satellite Surface Stress Working Group*, Nova University/NYIT Press, Fort Lauderdale, Florida, 153pp.

OMB (Office of Management and the Budget) 1992. *U.S. Actions for a Better Environment: A Sustained Commitment.* Washington, D.C., 44 pp.

Pacanowski, R. and S.G.H. Philander 1981. Parameterization of vertical mixing in numerical models of tropical oceans. *J. Geophys. Res.* **11**, 1443–1451.

Palmer, T.N. and D.A. Mansfield 1984. Response of two atmospheric general circulation models to sea-surface temperature anomalies in the tropical East and West Pacific. *Nature* **310**, 483–485.

Palmer, T. 1985. Response of the UK Meteorological Office general circulation model to sea-surface temperature anomalies in the tropical Pacific Ocean. In Nihoul 1985, 83–107.

Palmer, T.N., C. Brankovic, P. Viterbo, M.J. Miller 1992. Modeling interannual variations of summer monsoons. *J. Climate* **5**, 399–417.

Palmer, T.N. 1993. A nonlinear dynamical perspective on climate change. *Weather* **48**, 314–326.

Penland, C. and T. Magorian 1993. Prediction of NINO3 sea surface temperatures using linear inverse modeling. *J. Climate* **6**, 1067–1076.

Penland, C. and P.D. Sardeshmukh 1995. The optimal growth of tropical sea surface temperature anomalies. *J. Climate* **8**, 1999–2024.

Philander, S.G.H. and R.C. Pacanowski 1980. The generation of equatorial currents. *J. Geophys. Res.* **85**, 1123–1136.

Philander, S.G.H. 1981. The response of equatorial oceans to a relaxation of the trade winds. *J. Phys. Oceanogr.* **11**, 176–189.

Philander, S.G.H. and R.C. Pacanowski 1981a. Response of equatorial oceans to periodic forcing. *J. Geophys. Res.* **86**, 1903–1916.

Philander, S.G.H. and R.C. Pacanowski 1981b. The oceanic response to cross-equatorial winds (with applications to coastal upwelling in low latitudes). *Tellus* **33**, 201–210.

Philander, S.G.H., T. Yamagata, and R.C. Pacanowski 1984. Unstable air–sea interactions in the tropics. *J. Atmos. Sci.* **41**, 604–613.

Philander, S.G.H. and A.D. Seigel 1985. Simulation of El Niño of 1982–1983. In Nihoul 1985.

Philander, S.G.H., W. Hurlin, and A.D. Seigel 1987. A model of the seasonal cycle in the tropical Pacific Ocean. *J. Phys. Oceanogr.* **17**, 1986–2002.

Philander, S.G.H. 1990. *El Niño, La Niña, and the Southern Oscillation.* Academic Press, San Diego, 293pp.

Philander, S.G.H., R.C. Pacanowski, N.C. Lau, and M.J. Nath 1992. A simulation of the Southern Oscillation with a global atmospheric GCM coupled to a high-resolution, tropical Pacific Ocean GCM. *J. Climate* **5**, 308–329.

Picaut, J., A.J. Busalacchi, M.J. McPhaden, and B. Camusat 1990. Validation of the geostrophic method for estimating zonal currents at the equator from GEOSAT altimeter data. *J. Geophys. Res.* **95**, 3015–3024.

Prabhakara, C., D.A. Short, W. Wiscombe, and R.S. Fraser 1985. El Niño and atmospheric water vapor: Observations from Nimbus-7 SMMR. *J. Clim. Appl. Meteorol.* **24**, 1311–1324.

Ramanathan, V. and W. Collins 1991. Thermodynamic regulation of ocean warming by cirrus clouds deduced from observations of the 1987 El Niño. *Nature* **351**, 27–32.

Ramanathan, V., B. Subasilar, G.J. Zhang, W. Conant, R.D. Cess, J.T. Kiehl, H. Grassl, and L. Shi 1995. Warm pool heat budget and shortwave cloud forcing: A missing physics? *Science* **267**, 499–503.

Rasmusson, E.M. and T. Carpenter 1982. Variations in tropical sea surface temperature and surface wind fields associated with the Southern Oscillation / El Niño. *Mon. Weather Rev.* **110**, 354–384.

Rasmusson, E.M. and J.M. Wallace 1983. Meteorological aspects of the El Niño / Southern Oscillation. *Science* **222**, 1195–1202.

Rasmusson, E.M. and K. Mo 1993. Annual cycle of the global atmospheric water balance derived from NMC operational products. *Proceedings of the 4th Symposium on Global Change Studies,* Anaheim, California, 17–22 January 1993, American Meteorological Society, Boston, 345–350.

Rebert, J.P., J.-R. Donguy, G. Eldin, and K. Wyrtki 1985. Relations between sea level, thermocline depth, heat content, and dynamic height in the tropical Pacific Ocean. *J. Geophys. Res.* **90**, 11719–11725.

Reynolds, R.W. 1982. A monthly averaged climatology of sea surface temperature. NOAA T.R. NWS 31, U.S. Department of Commerce, 35pp.

Reynolds, R.W. and T.M. Smith 1994. Improved global sea surface temperature analyses using optimal interpolation. *J. Climate* **7**, 929–948.

Richardson, P.L. and W. Schmitz 1993. Deep cross-equatorial flow in the Atlantic measured with SOFAR floats. *J. Geophys. Res.* **98**, 8371–8387.

Robertson, A.W., C.-C. Ma, C.R. Mechoso, and M. Ghil 1995a. Simulation of the tropical Pacific climate with a coupled ocean–atmosphere general-circulation model. Part I: The seasonal cycle. *J. Climate* **8**, 1178–1198.

Robertson, A.W., C.-C. Ma, C.R. Mechoso, and M. Ghil 1995b. Simulation of the Tropical Pacific climate with a coupled ocean–atmosphere general circulation model. Part II: Interannual variability. *J. Climate* **8**, 1199–1216.

Rosati, A., K. Miyakoda, and R. Gudgel 1996. The impact of ocean initial conditions on ENSO forecasting with a coupled model. *Mon. Weather Rev.*, in press.

Rosenzweig, C. 1995. Agronomic effects of interannual climate variability/ENSO. In proceedings of a workshop on *Assessing the Vulnerability of Agriculture to Variations in Climate and Air Quality,* 19–21 June 1995, C. Rosenzweig, H. Wilson, and E.K Hartig (eds.), NASA/GISS Conference Publication, Goddard Institute for Space Studies, New York, 4.

Rossow, W.B. and R.A. Schiffer 1991. ISCCP cloud products. *Bull. Am. Meteor. Soc.* **72**, 2–20.

Rowntree, P.R. 1972. The influence of tropical east Pacific Ocean temperatures on the atmosphere. *Quart. J. Roy. Meteor. Soc.* **98**, 290–321.

Rowntree, P.R. 1976. Tropical forcing of atmospheric motions in a numerical model. *Quart. J. Roy. Meteor. Soc.* **102**, 583–605.

Sarachik, E.S. 1985. Modeling sea-surface temperature and its variability. In *Proceedings of the First National Workshop on the Global Weather Experiment*, National Academy Press, Washington, D.C., 765–778.

Sardeshmukh, P.D. and B.J. Hoskins 1988. The generation of global rotational flow by steady idealized tropical divergence. *J. Atmos. Sci.* **45**, 1228–1251.

Saucen, R., K. Barthel, and K. Hasselman 1988. Coupled ocean–atmosphere models with flux correction. *Climate Dynam.* **2**, 145–163.

Schneider, E.K. and J.L. Kinter 1994. An examination of internally generated variability in long climate simulations. *Climate Dynam.* **10**, 181–204.

Schopf, P.S. and M.J. Suarez 1988. Vacillations in a coupled ocean–atmosphere model. *J. Atmos. Sci.* **45**, 549–566.

Schott, F. and C.W. Boning 1991. The WOCE model in the western equatorial Atlantic: Upper-layer circulation. *J. Geophys. Res.* **96**, 6993–7004.

Seager, R., M.A. Cane, and S.E. Zebiak 1988. A model of the tropical Pacific sea surface temperature climatology. *J. Geophys. Res.* **93**, 1265–1280.

Seager, R. 1989. Modeling tropical Pacific sea surface temperature: 1970–1987. *J. Phys. Oceanogr.* **19**, 419–434.

SEQUAL 1982. The SEQUAL Wind Program. Lamont-Doherty Geological Observatory, Palisades, New York, 22pp.

Shen, G.T., L.J. Linn, T.M. Campbell, J.E. Cole, and R.G. Fairbanks 1992a. A chemical indicator of trade wind reversal in corals from the western tropical Pacific. *J. Geophys. Res.* **97**, 12689–12697.

Shen, G.T., J.E. Cole, D.W. Lea, L.J. Linn, T.A. McConnaughey, and R.G. Fairbanks 1992b. Surface ocean variability at Galapagos from 1936–1982: Calibration of geochemical tracers in corals. *Paleoceanography* **7**, 563–588.

Shinoda, T. 1993. Variation of the Thermohaline Structure in the Western Equatorial Pacific Upper Ocean. Ph.D. thesis, Univ. of Hawaii, 190 pp.

Shinoda, T. and R. Lukas 1995. Langrangian mixed layer modeling of the western equatorial Pacific. *J. Geophys. Res.* **100**, 2523–2541.

Shukla, J. and D.A. Paolino 1983. The Southern Oscillation and long range forecasting of the summer monsoon rainfall over India. *Mon. Weather Rev.* **111**, 1830–1837.

Shukla, J. and J.M. Wallace 1983. Numerical simulation of the atmospheric response to equatorial Pacific sea surface temperature anomalies. *J. Atmos. Sci.* **40**, 1613–1630.

Siegal, D.A., J.C. Ohlmann, L. Washburn, R.R. Bidigare, C.T. Nosse, E. Fields, and Y. Zhou 1995. Solar radiation, phytoplankton pigments, and the radiant heating of the equatorial warmpool. *J. Geophys. Res.* **100**, 4885–4891.

Siegenthaler, U. 1990. El Niño and atmospheric CO_2. *Nature* **345**, 295–296.

Simmons, A.J. 1982. The forcing of stationary wave motion by tropical diabatic heating. *Quart. J. Roy. Meteorol. Soc.* **108**, 503–534.

Simmons, A.J., J.M. Wallace, and G.W. Branstator 1983. Barotropic wave propagation and instability, and atmospheric teleconnection patterns. *J. Atmos. Sci.* **40**, 1363–1392.

Simpson, H.J., M.A. Cane, S.K. Lin, S.E. Zebiak, and A.L. Herczeg 1993. Forecasting annual discharge of River Murray, Australia, from a geophysical model of ENSO. *J. Climate* **6**, 386–90.

Slutz, R.J., S.J. Lubker, J.D. Hiscox, S.D. Woodruff, R.L. Jenne, D.H. Joseph, P.M. Steurer, and J.D. Elms 1985. Comprehensive Ocean-Atmopshere Data Set (COADS): Release 1. Cooperative Institute for Research in Environmental Science, Boulder, Colorado, 262pp.

Smith, T., R.W. Reynolds, and C.F. Ropelewski 1994. Optimal averaging of seasonal sea surface temperatures and associated confidence intervals. (1860–1989). *J. Climate* **7**, 949–964.

Smyth, W.D., D. Heber, and J.N. Moum 1996a. Local ocean response to a multiphase westerly windburst. Part 2: Thermal and freshwater responses. *J. Geophys. Res.* **100** (C10), 22 513–22 533.

Smyth, W.D., P.O. Zavialov, and J.N. Moum 1996b. Decay of turbulence in the upper ocean following sudden isolation from surface forcing. *J. Phys. Oceanogr.*, in press.

Soloviev, A. and R. Lukas 1996. Observation of large diurnal warming events in the near-surface layer of the western equatorial Pacific warm pool. *Deep-Sea Res.*, in press.

Stahle, D.W. and M.K. Cleaveland 1993. Southern Oscillation extremes reconstructed from tree rings of the Sierra Madre Occidental and southern Great Plains. *J. Climate* **6**, 129–140.

Stoffelen, A. and D.L.T. Anderson 1992. ERS-1 scatterometer calibration and validation activities at ECMWF: A. The quality and characteristics of the RADAR backscatter measurements. In *Proc. European "International Space Year" Conference*, Munich, Germany, 30 March to 4 April 1992.

von Storch, H. and H.A. Kruse 1985. The significant tropospheric midlatitudinal El Niño response patterns observed in January 1983 and simulated by a GCM. In Nihoul 1985, 275–288.

Streten, N.A. 1983. Extreme distributions of Australian rainfall in relation to sea surface temperature. *J. Climatol.* **3**, 143–153.

Suarez, M.J. and P.S. Schopf 1988. A delayed action oscillator for ENSO. *J. Atmos. Sci.* **45**, 3283–3287.

Subcommittee on Global Change Research 1995. *Our Changing Planet: The FY 1995 U.S. Global Change Research Program.* Office of Science and Technology Policy, Washington, D.C., 132pp.

Sybrandy, A.L. and P.P. Niiler 1991. WOCE/TOGA SVP Lagrangian Drifter Construction Manual. Scripps Institution of Oceanography Ref. 91/6, WOCE Report #63, 58pp.
TCIPO (TOGA COARE International Project Office) 1991. TOGA COARE Experimental Design. University Corporation for Atmospheric Research, Boulder, Colorado, 90 pp.
TCIPO 1992. TOGA COARE Operations Plan, Working Version. University Corporation for Atmospheric Research, Boulder, Colorado.
TCIPO 1993. TOGA COARE Intensive Observing Period Operations Summary. University Corporation for Atmospheric Research, Boulder, Colorado.
TCIPO 1995. Summary Report of the TOGA COARE International Data Workshop, Toulouse, France, 2–11 August 1994. University Corporation for Atmospheric Research, Boulder, Colorado, 170 pp.
Torrence, C. and P.J. Webster 1996. Low frequency variability and the annual cycle. In *Proceedings of the International TOGA Conference*, Melbourne, Australia, 1–7 April 1995, World Climate Research Program-91, WMO/TD No. 717, Geneva, 666–669.
Tourre, Y.M. and W.B. White 1996. Evolution of the ENSO signal over the Indo-Pacific domain. *J. Phys. Oceanogr.*, in press.
Trenberth, K.E. and D.A. Paolino 1981. Characteristic patterns of variability of sea level pressure in the Northern Hemisphere. *Mon. Weather Rev.* **109**, 1169–1189.
Trenberth, K.E. and D.J. Shea 1987. On the evolution of the Southern Oscillation. *Mon. Weather Rev.* **115**, 3078–3096.
Trenberth, K.E. and G.W. Branstator 1992. Issues in establishing causes of the 1988 drought over North America. *J. Climate* **5**, 159–172.
Trenberth, K.E. and J.W. Hurrell 1994. Decadal atmosphere–ocean variations in the Pacific. *Clim. Dynam.* **9**, 303–319.
Trenberth, K.E. 1995. El Niño / Southern Oscillation. In *Climate Change: Developing Southern Hemisphere Perspectives,* Thomas W. Giambelluca and Ann Henderson-Sellers (eds.), Wiley, New York, chapter 6.
Trenberth, K.E., and T.J. Hoar 1996. The 1990–1995 El Niño–Southern Oscillation event: Longest on record. *Geophys. Res. Lett.* **23**, 57–60.
Troup, A.J. 1965. The Southern Oscillation. *Quart. J. Roy. Meteorol. Soc.* **91**, 490–506.
Tsuchiya, M., R. Lukas, R.A. Fine, E. Firing, and E. Lindstrom 1989. Source waters of the Pacific Equatorial Undercurrent. *Prog. Oceanogr.* **23**, 101–147.
Tziperman, E., L. Stone, M. Cane, and H. Jarosh 1994. El Niño chaos: Overlapping of resonances between the seasonal cycle and the Pacific ocean–atmosphere oscillator. *Science* **264**, 72–74.

Wakata, Y. and E.S. Sarachik 1991a. On the role of equatorial ocean modes in the ENSO cycle. *J. Phys. Oceanogr.* **21**, 434–443.

Wakata, Y. and E.S. Sarachik 1991b. Unstable coupled atmosphere–ocean basin modes in the presence of a spatially varying basic state. *J. Atmos. Sci.* **48**, 2060–2077.

Wakata, Y. and E.S. Sarachik 1994. Nonlinear effects in coupled model atmosphere–ocean basin modes. *J. Atmos. Sci.* **51**, 909–920.

Waliser, D.E. 1996. Some considerations on the thermostat hypothesis. *Bull. Am. Meteorol. Soc.* **77**, 357–360.

Walker, G. T. 1924. Correlation in seasonal variations of weather IX: A further study of world weather. *Mem. Indian Meteor. Dept.* **24** (4), 275–332.

Walker, G T. and E.W. Bliss 1932. World Weather V. *Mem. Roy. Meteor. Soc.* **4** (36), 53–84.

Wang, B. and T. Li 1993. A simple atmsophere model of relevance to short term climate variations. *J. Atmos. Sci.* **50**, 260–284.

WCRP (World Climate Research Programme) 1983. *Large Scale Oceanographic Experiments in the World Climate Research Programme: Report of the JSC/CCCO Conference in Tokyo, 10–21 May 1983.* International Oceanographic Commission, Paris.

WCRP 1985. *Scientific Plan for the Tropical Ocean and Global Atmosphere Programme.* WMO/TD No. 64, World Meteorological Organization, Geneva, 146pp.

WCRP 1990. *Scientific Plan for the TOGA Coupled Ocean–Atmosphere Response Experiment.* WMO/TD No. 64 Addendum, World Meteorological Organization, Geneva.

WCRP 1992. *Simulation of Interannual and Intraseasonal Monsoon Variability.* WCRP-68, World Meteorological Organization, Geneva.

WCRP 1993. TOGA COARE International Participation Summaries: Status report April 1993. Appendix D of *WMO/IOC Inter-Governmental TOGA Board Report of the Sixth Session.* WMO/TD No. 559, World Meteorological Organization, Geneva, 12pp.

WCRP 1995. *CLIVAR—A Study of Climate Variability and Predictability.* WCRP-89, WMO/TD No. 690, World Meteorological Organization, Geneva, 157pp.

Weare, B.C., A. Navato, and R.E. Newell 1976. Empirical orthogonal analysis of Pacific sea surface temperatures. *J. Phys. Oceanogr.* **6**, 671–678.

Weaver, C.P., W.P. Collins, and H. Grassl 1994. The relationship between clear-sky atmospheric greenhouse effect and deep convection during the Central Equatorial Pacific Experiment (CEPEX): Model calculations and satellite observations. *J. Geophys. Res.* **99** (D12), 25891–25901.

Webster, P.J. 1972. Response of the tropical atmosphere to local steady forcing. *Mon. Weather Rev.* **100**, 518–541.

Webster, P. J. 1981. Mechanisms determining the atmospheric response to sea surface temperature anomalies. *J. Atmos. Sci.* **38**, 554–571.

Webster, P. J. 1982. Seasonality in the local and remote atmospheric response to sea surface temperature anomalies. *J. Atmos. Sci.* **39**, 41–52.

Webster, P.J. and R.A. Houze, Jr. 1991. The Equatorial Mesoscale Experiment (EMEX): An overview. *Bull. Am. Meteorol. Soc.* **72**, 1481–1505.

Webster, P.J. and R. Lukas 1992. The Tropical Ocean / Global Atmosphere Coupled Ocean–Atmosphere Response Experiment (COARE). *Bull. Am. Meteorol. Soc.* **73**, 1377–1416.

Webster, P.J. and S. Yang 1992. Monsoon and ENSO: Selectively interactive systems. *Quart. J. Roy. Meteorol. Soc.* **118**, 877–926.

Webster, P.J., C.A. Clayson, and J.A. Curry 1996. Clouds, radiation and the diurnal cycle of sea surface temperature in the tropical western Pacific. *J. Climate* **9**, 1712–1730.

Weller, R.A., and S.P. Anderson 1996. Surface meteorology and air-sea fluxes in the western equatorial Pacific warm pool during the TOGA Coupled Ocean–Atmosphere Response Experiment. *J. Climate* **9**, 1959–1990.

Wentz, F.J. 1989. User's manual: SSMI geophysical tapes. RSS Technical Report 060980, Remote Sensing Systems, Santa Rosa, California, 16pp.

Wentz, F.J. 1992. Measurement of oceanic wind vector using satellite radiometers. *IEEE Trans. Geosci. Rem. Sens.* **30**, 960–972.

Wilson, D., E. Johns, and R.L. Molinari 1994. Upper layer circulation in the western tropical North Atlantic Ocean during August 1989. *J. Geophys. Res.* **99** (C11), 22513–22523.

Winguth, A.M, M. Heimann, K.D. Kurz, E. Maier-Reimer, U. Mikolajewicz, and J. Segschneider 1994. El Niño-Southern Oscillation related fluctuations of the marine carbon cycle. *Global Biogeochem. Cycles* **8**, 39–63

WMO (World Meteorological Organization) 1990. *Global Precipitation Climatology Project: Implementation and Data Management Plan.* WMO/TD No. 367, World Meteorological Organization, Geneva.

Wong, C-S., Y.-H. Chan, J.S. Page, G.E. Smith, and R.D. Bellegay 1993. Changes in the equatorial CO_2 flux and new production estimated from CO_2 and nutrient levels in Pacific surface waters during the 1986–87 El Niño. *Tellus* **45B**, 64–79.

Wyrtki, K. 1973. Teleconnections in the equatorial Pacific Ocean. *Science* **180**, 66–68.

Wyrtki, K. 1974. Equatorial currents in the Pacific 1950 to 1970 and their relations to the trade winds. *J. Phys. Oceanogr.* **4**, 372–380.

Wyrtki, K. 1975. El Niño—The dynamic response of the equatorial Pacific Ocean to atmospheric forcing. *J. Phys. Oceanogr.* **5**, 572–584.

Wyrtki, K. and G. Meyers 1975. The trade wind field over the Pacific Ocean. II. Bimonthly fields of wind stress 1950–1972. Hawaii Institute of Geophysics Report HIG 75-2, University of Hawaii, Honolulu.

Wyrtki, K., E. Stroup, W. Patzert, R. Williams, and W. Quinn 1976. Predicting and observing El Niño. *Science* **191**, 343–346.

Wyrtki, K. 1978. Lateral oscillations of the Pacific Equatorial Countercurrent. *J. Phys. Oceanogr.* **8**, 530–532.

Wyrkti, K. 1979. The response of sea surface topography to the 1976 El Niño. *J. Phys. Oceanogr.* **9**, 1223–1231.

Wyrtki, K. 1985. Water displacements in the Pacific and the genesis of El Niño cycles. *J. Geophys. Res.* **90**, 7129–7132.

Wyrtki, K. 1987. Indices of equatorial currents in the central Pacific. *Trop. Ocean–Atmos. Newslett.* **38**, 3–5.

Xie, S.P. 1995. Interaction between the annual and interannual variations in the equatorial Pacific. *J. Phys. Oceanogr.* **25**, 1930–1941.

Xu, J.-S. and H. von Storch 1990. Principal oscillation patterns—Prediction of the state of ENSO. *J. Climate* **3**, 1316–1329.

Yang, S. 1996. ENSO–snow–monsoon associations and seasonal-to-interannual predictions. *Int. J. Climate* **16**, 125–134.

Yin, F.L., and E.S. Sarachik 1993. Dynamics and heat balance of steady equatorial undercurrents. *J. Phys. Oceanogr.* **23**, 1647–1669.

Young, G.S., S.M. Perugiui, and C.W. Fairall 1995. Convective wakes in the equatorial western Pacific during TOGA. *Mon. Weather Rev.* **123**, 110–123.

Zebiak, S.E. 1982. A simple atmospheric model of relevance to El Niño. *J. Atmos. Sci.* **39**, 2017–2027.

Zebiak, S.E. 1984. Tropical Atmosphere–Ocean Interaction and the El Niño / Southern Oscillation Phenomenon. Ph.D. thesis, Massachusetts Institute of Technology, 261pp.

Zebiak, S.E. 1986. Atmospheric convergence feedback in a simple model for El Niño. *Mon. Weather Rev.* **114**, 1263–1271.

Zebiak, S.E. and M.A. Cane 1987. A model El Niño / Southern Oscillation. *Mon. Weather Rev.* **115**, 2262–2278.

Zebiak, S.E. 1989. On the 30–60 day oscillation and the prediction of El Niño. *J. Climate* **2**, 1381–1387.

Zhang, G.J., and M.J. McPhaden 1995. The relationship between sea surface temperature and latent heat flux in the equatorial Pacific. *J. Climate* **8**, 589–605.

Zhang, Y., J.M. Wallace, and D.S. Battisti 1996. ENSO-like decade-to-century scale variability: 1900–93. *J. Climate*, in press.

APPENDICES

A. MEMBERS OF THE TOGA PANEL

Richard T. Barber, 1987–90
Tim P. Barnett, 1984–86
Maurice Blackmon, 1987–90
Otis B. Brown, 1985–91
Antonio J. Busalacchi, 1989–96
Mark A. Cane, 1984–89
Robert E. Dickinson, 1994–96
Steven Esbensen, 1988–96
Rana A. Fine, 1990–94
David Halpern, 1990–96
D. Edmunds Harrison, 1988–90
Dennis L. Hartmann, 1990–96
Paul R. Julian, 1986–89
Eli J. Katz, 1984–86
Robert A. Knox, 1991–96
Ants Leetmaa, 1984–87, 1992–96
Roger Lukas, 1988–96
Pearn P. Niiler, 1987–90
S. George H. Philander, 1984–87
Eugene M. Rasmusson, 1984–86
Edward S. Sarachik, 1991–96, Chair 1992–96
Jagadish Shukla, 1984–92, Chair 1989–92
Joanne Simpson, 1991–94
Kevin E. Trenberth, 1984–87
John M. Wallace, 1985–90, Chair 1985–89
Ferris Webster, 1984–85, Chair 1984
Peter J. Webster, 1984–86
John A. Young, 1985–91
Stephen E. Zebiak, 1990–96

B. TOGA PRODUCTS

TOGA CD-ROMs. The 1994 edition consisted of 1985–1990 TOGA data. It included: sea surface temperature, sub-surface thermal data, surface marine observations, TAO data, pseudo-stress fields, cloud data, precipitation data, sea levels, and associated ECMWF analyses. The 6-CD package and Users' Guide available from:
    User Services Office, MS300/320
    JPL PO.DAAC
    Jet Propulsion Laboratory
    4800 Oak Grove Drive
    Pasadena, CA 91109.

Climate Diagnostics Bulletin. Published monthly since 1982. Contains: (1) monthly means and anomalies in the tropics of sea surface temperature, outgoing long-wave radiation, winds, pressure, susbsurface ocean thermal structure, and sea level; (2) Forecast Forum; and (3) monthly means and anomalies in the extratropics of sea-level pressure, height of the 500-mb surface, and indices of "teleconnections". Available from:
    Climate Prediction Center
    Attn: Climate Diagnostics Bulletin
    NOAA/NWS/NCEP
    World Weather Building, Room 605
    5200 Auth Road
    Washington, DC 20233.
The bulletin is also available on the World Wide Web from site (http://nic.fb4.noaa.gov).

Experimental Long Lead Forecast Bulletin. Published quarterly since 1992. Contains: experimental forecasts of tropical sea surface temperature, Southern Oscillation Index, rainfall, Atlantic storm activity, U.S. surface conditions and precipitation, and South African rainfall. Available from:
    Climate Prediction Center, W/NMC51
    Attn: Experimental Long Lead Bulletin
    NOAA/NWS/NCEP
    World Weather Building
    5200 Auth Road
    Washington, DC 20233.

Monitor Climático. Published monthly since 1987 in Portugese. Contains: monthly precipitation in northeast Brazil, monthly position of the Atlantic Inter-Tropical Convergence Zone, monthly means and anomalies of tropical sea surface temperature, monthly means and anomalies of tropical sea-level pressure, monthly mean and anomalies of outgoing long-wave radiation, and monthly means and anomalies of winds at heights of both 850 mb and 200 mb. Available from:
>Divisão de Tempo e Clima
>Fundação Cearense de Meteorologia e Recursos Hídrocos—FUNCEME
>Av. Bezerra de Menezes, No 1.900—São Gerardo
>Caixa Postal D-3221
>Fortaleza, BRAZIL.

World Wide Web Products. A number of products have become available on the World Wide Web. Uniform Resource Locators (URLs) and descriptions are listed below. Web sites are subject to frequent change.

•Climate Prediction Center of the National Centers for Environmental Prediction site (http://nic.fb4.noaa.gov) provides prediction products and data products relevant to ENSO and to its effects on U.S. climate.

•Lamont-Doherty Climate Group site (http://rainbow.ldgo.columbia.edu) provides data products, ENSO advisories, and forecasts of ENSO using the Zebiak-Cane model.

•NOAA/Pacific Marine Environment Laboratory (PMEL) site (http://www.pmel.noaa.gov/toga-tao/realtime.html) points to the latest data on surface and subsurface conditions in the tropical Pacific. It also provides tools for obtaining past TOGA TAO data. Also available from PMEL is a site (http://www.pmel.gov/toga-tao/el-nino/home.html) that gives a good introduction to ENSO and its effects. The site also points to ENSO prediction information.

•Global Change Master Directory site (http://gcmd.gsfc.nasa.gov) guides users to climate data, including satellite data.

•National Climatic Data Center site (http://www.ncdc.gov) contains guides to online data services and climatological products. It also points to the Climate Variations Bulletin, which documents monthly climate anomalies, mostly over the United States.

•Center for Ocean–Land–Atmosphere Studies site (http:/grads.iges.org/home.html) points to coupled-model forecasts of ENSO and other related sites.

- University of Hawaii School of Ocean and Earth Science and Technology site (http://www.soest.hawaii.edu) provides data from the Hawaii Sea Level Data Center and points to other online data centers.
- NOAA Seasonal-to-Interannual site (http://www.noaa.gov/seasonal.html) gives NOAA's plans and rationale for predictions of seasonal-to-interannual climate variations.
- NOAA Office of Global Programs site (http://www.noaa.gov/ogpl) points to El Niño information and the status of the International Research Institute (for Climate Prediction).
- Florida State University site (http://coaps.fsu.edu) provides a guide to a number of meteorological and oceanographic servers.
- University of Washington site (http://www.atmos.washington.edu) (choose "Local Resources") provides precipitation climatologies of the tropics, a T-POP reference list, and pointers to other meteorological servers.

C. ACRONYMS AND OTHER ABBREVIATIONS

ADCP	acoustic Doppler current profiler
ADEOS	Advanced Earth Observing Satellite (Japan)
AIREP	Aircraft En-Route Report
AMEX	Australian Monsoon Experiment
AOML	Atlantic Oceanographic and Meteorological Laboratory (NOAA)
ARGOS	ARGOS Data Collection and Platform Location System (provided by France)
ATLAS	Autonomous Temperature Line Acquisition System
ATM	Atmospheric Sciences Division (NSF)
AVHRR	Advanced Very-High Resolution Radiometer
BSN	Basic Synoptic Network (of the World Weather Watch)
C	carbon
C&GC	Climate and Global Change (NOAA)
CAC	Climate Analysis Center, now CPC (NOAA)
CCCO	Committee for Climatic Change and the Ocean
CEPEX	Central Pacific Experiment
CLIVAR	Study of Climate Variability and Predictability (WCRP)
CO_2	carbon dioxide
COARE	Coupled Ocean–Atmosphere Response Experiment
CPC	Climate Prediction Center (NOAA)
CRC	Climate Research Committee (NRC)
CZCS	Coastal Zone Color Scanner
DMSP	Defense Meteorological Satellite Program
DOE	Department of Energy
ECMWF	European Centre for Medium-Range Weather Forecasting
EMEX	Equatorial Monsoon Experiment
ENSO	El Niño and the Southern Oscillation
EOF	empirical orthogonal function
EPOCS	Equatorial Pacific Ocean Climate Studies
EqPac	Equatorial Pacific Experiment
ERS	Earth Resources Satellite (ESA)
ESA	European Space Agency
EUC	Equatorial Undercurrent
FOCAL	Français Océan Climat Atlantique Equatorial (SEQUAL/FOCAL)
FGGE	First GARP Global Experiment
FUNCEME	Ceara Foundation for Meteorology and Hydrological Resources (Northeast Brazil)
ftp	file-transfer protocol

GARP	Global Atmospheric Research Program
GATE	GARP Atlantic Tropical Experiment
GCM	general-circulation model
GCOS	Global Climate Observing System
Geosat	Geodesy satellite (U.S. Navy)
GFDL	Geophysical Fluid Dynamics Laboratory (NOAA)
GISS	Goddard Institute for Space Studies (NASA)
GOALS	Global Ocean–Atmosphere–Land System program
GOES	Geostationary Operational Environmental Satellite
GOOS	Global Ocean Observing System
GPCP	Global Precipitation Climatology Project
GTS	Global Telecommunication System
ICSU	International Council of Scientific Unions
IFA	Intensive Flux Array
INDEX	Indian Ocean Experiment
IOC	Intergovernmental Oceanographic Commission
IOP	intensive observation period
IRICP	International Research Institute for Climate Prediction
ITPO	International TOGA Project Office
IUGG	International Union of Geodesy and Geophysics
ITCZ	Intertropical Convergence Zone
JGOFS	Joint Global Ocean Flux Study
JMA	Japan Meteorological Agency
JOC	Joint Organizing Committee (WCRP)
JPL	Jet Propulsion Laboratory
JSC	Joint Scientific Committee (WCRP)
LIA	Line Islands Array
MDCRS	Meteorological Data Collection and Reporting Systems
MONEG	Monsoon Experimentation Group
MTPE	Mission to Planet Earth (NASA)
NASA	National Aeronautics and Space Administration
NCEP	National Centers for Environmental Prediction (NOAA)
NECC	North Equatorial Countercurrent
NMC	National Meteorological Center, now NCEP (NOAA)
NOAA	National Oceanic and Atmospheric Administration
NORPAX	North Pacific Experiment
NOS	National Ocean Service
NRC	National Research Council
NROSS	Navy Remote Ocean Sensing System
NSCAT	NASA Scatterometer
NSF	National Science Foundation
OCE	Ocean Sciences Division (NSF)

OLR	outgoing long-wave radiation
OMB	Office of Management and the Budget
ONR	Office of Naval Research
OSSE	observing-system simulation experiment
pCO_2	partial pressure of carbon dioxide
PEQUOD	Pacific Equatorial Ocean Dynamics Experiment
PMEL	Pacific Marine Environmental Laboratory
ppmv	parts per million by volume
SEQUAL	Seasonal Equatorial Atlantic Experiment (SEQUAL/FOCAL)
SCPP	Seasonal-to-Interannual Climate Prediction Program (NOAA)
SIO	Scripps Institution of Oceanography
SMMR	Scanning Multichannel Microwave Radiometer
SOI	Southern Oscillation Index
SSG	Scientific Steering Group
SSM/I	Special Sensor Microwave/Imager
Sv	sverdrup, $10^6 m^3/s$
SWG	science working group
TAO	Tropical Atmosphere/Ocean
TCIPO	TOGA COARE International Project Office
TIROS	Television Infrared Observation Satellite
TIWE	Tropical Instability Wave Experiment
TOGA	Tropical Oceans and Global Atmosphere Program
TOPEX/ Poseidon	Ocean Topography Experiment (Poseidon is the French component of the experiment)
T-POP	TOGA Program on Prediction
TSLC	TOGA Sea Level Center
U.N.	United Nations
UNCED	U.N. Conference on Environment and Development
UNESCO	U.N. Educational, Scientific, and Cultural Organization
USGCRP	U.S. Global Change Research Program
VACM	vector-averaging current meter
VMCM	vector-measuring current meter
VOS	volunteer observing ship
WEPOCS	Western Equatorial Pacific Ocean Circulation Study
WESTRAX	Western Tropical Atlantic Experiment
WOCE	World Ocean Circulation Experiment
WCRP	World Climate Research Programme
WMO	World Meteorological Organization
WWW	World Weather Watch (WMO)
XBT	expendable bathythermograph
XCTD	expendable conductivity-temperature-depth probe